Lectures in Mathematics
ETH Zürich
Department of Mathematics
Research Institute of Mathematics

Managing Editor:
Oscar E. Lanford

Randall J. LeVeque

Numerical Methods for Conservation Laws

Second Edition

Springer Basel AG

Author's address:

Randall J. LeVeque
Department of Mathematics, GN-50
University of Washington
Seattle, WA 98195
USA

A CIP catalogue record for this book is available from the
Library of Congress, Washington D.C., USA

Deutsche Bibliothek Cataloging-in-Publication Data
LeVeque, Randall J.:
Numerical methods for conservation laws / Randall J. LeVeque.
Springer Basel AG, 1992
   (Lectures in mathematics : ETH Zürich)
    ISBN 978-3-7643-2723-1   ISBN 978-3-0348-8629-1 (eBook)
    DOI 10.1007/978-3-0348-8629-1

ISBN 978-3-7643-2723-1

9 8 7 6

Preface

These notes developed from a course on the numerical solution of conservation laws first taught at the University of Washington in the fall of 1988 and then at ETH during the following spring.

The overall emphasis is on studying the mathematical tools that are essential in developing, analyzing, and successfully using numerical methods for nonlinear systems of conservation laws, particularly for problems involving shock waves. A reasonable understanding of the mathematical structure of these equations and their solutions is first required, and Part I of these notes deals with this theory. Part II deals more directly with numerical methods, again with the emphasis on general tools that are of broad use. I have stressed the underlying ideas used in various classes of methods rather than presenting the most sophisticated methods in great detail. My aim was to provide a sufficient background that students could then approach the current research literature with the necessary tools and understanding.

Without the wonders of TeX and LaTeX, these notes would never have been put together. The professional-looking results perhaps obscure the fact that these are indeed lecture notes. Some sections have been reworked several times by now, but others are still preliminary. I can only hope that the errors are not too blatant. Moreover, the breadth and depth of coverage was limited by the length of these courses, and some parts are rather sketchy. I do have hopes of eventually expanding these notes into a full-fledged book, going more deeply into some areas, discussing a wider variety of methods and techniques, and including discussions of more applications areas. For this reason I am particularly interested in receiving corrections, comments and suggestions. I can be reached via electronic mail at *na.rleveque@na-net.stanford.edu*.

I am indebted to Jürgen Moser and the Forschungsinstitut at ETH for the opportunity to visit and spend time developing these notes, and to Martin Gutknecht for initiating this contact. During the course of this project, I was also supported in part by a Presidential Young Investigator Award from the National Science Foundation.

Contents

Part I

Mathematical Theory

1 Introduction

1.1 Conservation laws

These notes concern the solution of hyperbolic systems of conservation laws. These are time-dependent systems of partial differential equations (usually nonlinear) with a particularly simple structure. In one space dimension the equations take the form

$$\frac{\partial}{\partial t}u(x,t) + \frac{\partial}{\partial x}f(u(x,t)) = 0. \tag{1.1}$$

Here $u : \mathbb{R} \times \mathbb{R} \to \mathbb{R}^m$ is an m-dimensional vector of conserved quantities, or state variables, such as mass, momentum, and energy in a fluid dynamics problem. More properly, u_j is the density function for the jth state variable, with the interpretation that $\int_{x_1}^{x_2} u_j(x,t)\,dx$ is the total quantity of this state variable in the interval $[x_1, x_2]$ at time t.

The fact that these state variables are conserved means that $\int_{-\infty}^{\infty} u_j(x,t)\,dx$ should be constant with respect to t. The functions u_j themselves, representing the spatial distribution of the state variables at time t, will generally change as time evolves. The main assumption underlying (1.1) is that knowing the value of $u(x,t)$ at a given point and time allows us to determine the rate of flow, or **flux**, of each state variable at (x,t). The flux of the jth component is given by some function $f_j(u(x,t))$. The vector-valued function $f(u)$ with jth component $f_j(u)$ is called the **flux function** for the system of conservation laws, so $f : \mathbb{R}^m \to \mathbb{R}^m$. The derivation of the equation (1.1) from physical principles will be illustrated in the next chapter.

The equation (1.1) must be augmented by some initial conditions and also possibly boundary conditions on a bounded spatial domain. The simplest problem is the pure initial value problem, or **Cauchy problem**, in which (1.1) holds for $-\infty < x < \infty$ and $t \geq 0$. In this case we must specify initial conditions only,

$$u(x,0) = u_0(x), \qquad -\infty < x < \infty. \tag{1.2}$$

We assume that the system (1.1) is **hyperbolic**. This means that the $m \times m$ Jacobian matrix $f'(u)$ of the flux function has the following property: For each value of u the

1

eigenvalues of $f'(u)$ are real, and the matrix is diagonalizable, *i.e.*, there is a complete set of m linearly independent eigenvectors. The importance of this assumption will be seen later.

In two space dimensions a system of conservation laws takes the form

$$\frac{\partial}{\partial t}u(x,y,t) + \frac{\partial}{\partial x}f(u(x,y,t)) + \frac{\partial}{\partial y}g(u(x,y,t)) = 0 \qquad (1.3)$$

where $u : \mathbb{R}^2 \times \mathbb{R} \to \mathbb{R}^m$ and there are now two flux functions $f, g : \mathbb{R}^m \to \mathbb{R}^m$. The generalization to more dimensions should be clear.

Hyperbolicity now requires that any real linear combination $\alpha f'(u) + \beta g'(u)$ of the flux Jacobians should be diagonalizable with real eigenvalues.

For brevity throughout these notes, partial derivatives will usually be denoted by subscripts. Equation (1.3), for example, will be written as

$$u_t + f(u)_x + g(u)_y = 0. \qquad (1.4)$$

Typically the flux functions are nonlinear functions of u, leading to nonlinear systems of partial differential equations (PDEs). In general it is not possible to derive exact solutions to these equations, and hence the need to devise and study numerical methods for their approximate solution. Of course the same is true more generally for any nonlinear PDE, and to some extent the general theory of numerical methods for nonlinear PDEs applies in particular to systems of conservation laws. However, there are several reasons for studying this particular class of equations on their own in some depth:

- Many practical problems in science and engineering involve conserved quantities and lead to PDEs of this class.

- There are special difficulties associated with solving these systems (e.g. shock formation) that are not seen elsewhere and must be dealt with carefully in developing numerical methods. Methods based on naive finite difference approximations may work well for smooth solutions but can give disastrous results when discontinuities are present.

- Although few exact solutions are known, a great deal *is* known about the mathematical structure of these equations and their solution. This theory can be exploited to develop special methods that overcome some of the numerical difficulties encountered with a more naive approach.

1.2 Applications

One system of conservation laws of particular importance is the **Euler equations** of gas dynamics. More generally, the fundamental equations of fluid dynamics are the Navier-Stokes equations, but these include the effects of fluid viscosity and the resulting flux function depends not only on the state variables but also on their gradients, so the equations are not of the form (1.1) and are not hyperbolic. A gas, however, is sufficiently dilute that viscosity can often be ignored. Dropping these terms gives a hyperbolic system of conservation laws with $m = d + 2$ equations in d space dimensions, corresponding to the conservation of mass, energy, and the momentum in each direction. In one space dimension, these equations take the form

$$\frac{\partial}{\partial t}\begin{bmatrix} \rho \\ \rho v \\ E \end{bmatrix} + \frac{\partial}{\partial x}\begin{bmatrix} \rho v \\ \rho v^2 + p \\ v(E + p) \end{bmatrix} = 0, \tag{1.5}$$

where $\rho = \rho(x, t)$ is the density, v is the velocity, ρv is the momentum, E is the energy, and p is the pressure. The pressure p is given by a known function of the other state variables (the specific functional relation depends on the gas and is called the "equation of state"). The derivation of these equations is discussed in more detail in Chapters 2 and 5.

These equations, and some simplified versions, will be used as examples throughout these notes. Although there are many other systems of conservation laws that are important in various applications (some examples are mentioned below), the Euler equations play a special role. Much of the theory of conservation laws was developed with these equations in mind and many numerical methods were developed specifically for this system. So, although the theory and methods are applicable much more widely, a good knowledge of the Euler equations is required in order to read much of the available literature and benefit from these developments. For this reason, I urge you to familiarize yourself with these equations even if your primary interest is far from gas dynamics.

The shock tube problem. A simple example that illustrates the interesting behavior of solutions to conservation laws is the "shock tube problem" of gas dynamics. The physical set-up is a tube filled with gas, initially divided by a membrane into two sections. The gas has a higher density and pressure in one half of the tube than in the other half, with zero velocity everywhere. At time $t = 0$, the membrane is suddenly removed or broken, and the gas allowed to flow. We expect a net motion in the direction of lower pressure. Assuming the flow is uniform across the tube, there is variation in only one direction and the one-dimensional Euler equations apply.

The structure of this flow turns out to be very interesting, involving three distinct waves separating regions in which the state variables are constant. Across two of these waves there are discontinuities in some of the state variables. A **shock wave** propagates into the region of lower pressure, across which the density and pressure jump to higher values and all of the state variables are discontinuous. This is followed by a **contact discontinuity**, across which the density is again discontinuous but the velocity and pressure are constant. The third wave moves in the opposite direction and has a very different

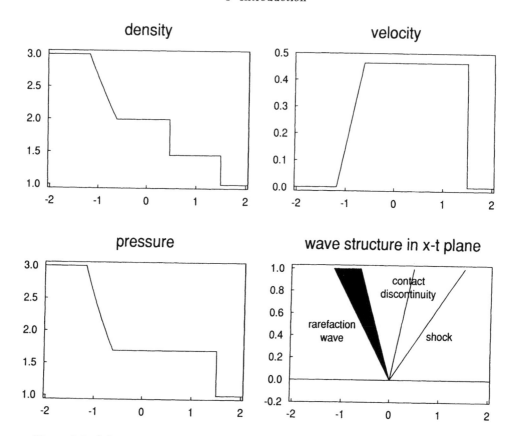

Figure 1.1. Solution to a shock tube problem for the one-dimensional Euler equations.

structure: all of the state variables are continuous and there is a smooth transition. This wave is called a **rarefaction wave** since the density of the gas decreases (the gas is rarefied) as this wave passes through.

If we put the initial discontinuity at $x = 0$, then the resulting solution $u(x,t)$ is a "similarity solution" in the variable x/t, meaning that $u(x,t)$ can be expressed as a function of x/t alone, say $u(x,t) = w(x/t)$. It follows that $u(x,t) = u(\alpha x, \alpha t)$ for any $\alpha > 0$, so the solution at two different times t and αt look the same if we rescale the x-axis. This also means that the waves move at constant speed and the solution $u(x,t)$ is constant along any ray $x/t = $ constant in the x-t plane.

Figure 1.1 shows a typical solution as a function of x/t. We can view this as a plot of the solution as a function of x at time $t = 1$, for example. The structure of the solution in the x-t plane is also shown.

In a real experimental shock tube, the state variables would not be discontinuous

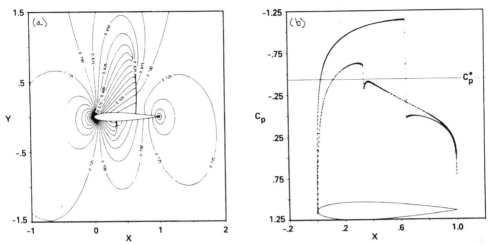

Figure 1.2. (a) Pressure contours for flow around an airfoil. (b) Pressure coefficient plotted along the top and bottom surface. Figure taken from Yee and Harten[100]. (Reprinted with permission.)

across the shock wave or contact discontinuity because of effects such as viscosity and heat conduction. These are ignored in the Euler equations. If we include these effects, using the full Navier-Stokes equations, then the solution of the partial differential equations would also be smooth. However, these smooth solutions would be nearly discontinuous, in the sense that the rise in density would occur over a distance that is microscopic compared to the natural length scale of the shock tube. If we plotted the smooth solutions they would look indistinguishable from the discontinuous plots shown in Figure 1.1. For this reason we would like to ignore these viscous terms altogether and work with the simpler Euler equations.

The Euler equations are used extensively in aerodynamics, for example in modeling the flow of air around an aircraft or other vehicle. These are typically three dimensional problems, although 2D and even 1D problems are sometimes of interest. A typical 2D problem is the flow of air over an airfoil, which is simply the cross section of a wing. Figure 1.2a (taken from Yee and Harten[100]) shows the contours of pressure in a steady state solution for a particular airfoil shape when the freestream velocity is Mach 0.8. Note the region above the upper surface of the airfoil where many contour lines coincide. This is again a shock wave, visible as a discontinuity in the pressure. A weaker shock is visible on the lower surface of the airfoil as well.

Small changes in the shape of an airfoil can lead to very different flow patterns, and so the ability to experiment by performing calculations with a wide variety of shapes is required. Of particular interest to the aerodynamical engineer is the pressure distribution along the airfoil surface. From this she can calculate the lift and drag (the vertical and horizontal components of the net force on the wing) which are crucial in evaluating its performance. Figure 1.2b shows the "pressure coefficient" along the upper and lower surfaces. Again the shocks can be observed as discontinuities in pressure.

The location and strength of shock waves has a significant impact on the overall solution, and so an accurate computation of discontinuities in the flow field is of great importance.

The flow field shown in Figure 1.2 is a **steady state** solution, meaning the state variables $u(x, y, t)$ are independent of t. This simplifies the equations since the time derivative terms drop out and (1.3) becomes

$$f(u)_x + g(u)_y = 0. \tag{1.6}$$

In these notes we are concerned primarily with time-dependent problems. One way to solve the steady state equation (1.6) is to choose some initial conditions (e.g. uniform flow) and solve the time-dependent equations until a steady state is reached. This can be viewed as an iterative method for solving the steady state equation. Unfortunately, it is typically a very inefficient method since it may take thousands of time steps to reach steady state. A wide variety of techniques have been developed to accelerate this convergence to steady state by giving up time accuracy. The study of such acceleration techniques is a whole subject in its own right and will not be presented here. However, the discrete difference equations modeling (1.6) that are solved by such an iterative method must again be designed to accurately capture discontinuities in the flow, and are often identical to the spatial terms in a time-accurate method. Hence much of the theory developed here is also directly applicable in solving steady state equations.

Unsteady problems also arise in aerodynamics, for example in modeling wing flutter, or the flow patterns around rotating helicopter blades or the blades of a turbine. At high speeds these problems involve the generation of shock waves, and their propagation and interaction with other shocks or objects is of interest.

Meteorology and weather prediction is another area of fluid dynamics where conservation laws apply. Weather fronts are essentially shock waves — "discontinuities" in pressure and temperature. However, the scales involved are vastly greater than in the shock tube or airfoil problems discussed above, and the viscous and dissipative effects cause these fronts to have a width of several miles rather than the fractions of an inch common in aerodynamics.

Astrophysical modeling leads to systems of conservation laws similar to the Euler equations for the density of matter in space. A spiral galaxy, for example, may consist of alternating arms of high density and low density, separated by "discontinuities" that are again propagating shock waves. In this context the shock width may be two or three light years! However, since the diameter of a galaxy is on the order of 10^5 light years, this is

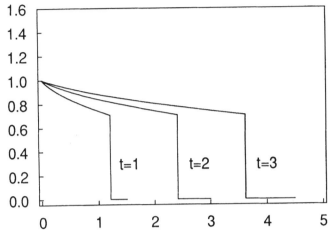

Figure 1.3. Solution of the Buckley-Leverett equation at three different times.

still a small distance in relative terms. In particular, in a practical numerical calculation the shock width may well be less than the mesh width.

Modeling the dynamics of a single star, or the plasma in a fusion reactor, also requires conservation laws. These now involve electromagnetic effects as well as fluid dynamics. The magnetohydrodynamics (MHD) equations are one system of this type.

Multiphase flow problems in porous materials give rise to somewhat different systems of conservation laws. One important application area is secondary oil recovery, in which water (with some additives, perhaps) is pumped down one well in an effort to force more oil out of other wells. One particularly simple model is the Buckley-Leverett equation, a scalar conservation law for a single variable u representing the saturation of water in the rock or sand ($u = 0$ corresponds to pure oil, $u = 1$ to pure water). Figure 1.3 shows the solution to this 1D problem at three different times. Here the initial condition is $u(x,0) = 0$ and a boundary condition $u(0,t) = 1$ is applied, corresponding to pure water being pumped in at the left boundary. Note that the advancing front again has a discontinuity, or propagating shock wave. The Buckley-Leverett equation is discussed further in Chapter 4. More realistic models naturally involve systems of conservation laws in two or three dimensions.

Systems of conservation laws naturally arise in a wide variety of other applications as well. Examples include the study of explosions and blast waves, the propagation of waves in elastic solids, the flow of glaciers, and the separation of chemical species by chromatography, to name but a few.

1.3   Mathematical difficulties

Discontinuous solutions of the type shown above clearly do not satisfy the PDE in the classical sense at all points, since the derivatives are not defined at discontinuities. We need to define what we mean by a solution to the conservation law in this case. To find the correct approach we must first understand the derivation of conservation laws from physical principles. We will see in Chapter 2 that this leads first to an integral form of the conservation law, and that the differential equation is derived from this only by imposing additional smoothness assumptions on the solution. The crucial fact is that the integral form continues to be valid even for discontinuous solutions.

Unfortunately the integral form is more difficult to work with than the differential equation, especially when it comes to discretization. Since the PDE continues to hold except at discontinuities, another approach is to supplement the differential equations by additional "jump conditions" that must be satisfied across discontinuities. These can be derived by again appealing to the integral form.

To avoid the necessity of explicitly imposing these conditions, we will also introduce the "weak form" of the differential equations. This again involves integrals and allows discontinuous solutions but is easier to work with than the original integral form of the conservation laws. The weak form will be fundamental in the development and analysis of numerical methods.

Another mathematical difficulty that we must face is the possible nonuniqueness of solutions. Often there is more than one weak solution to the conservation law with the same initial data. If our conservation law is to model the real world then clearly only one of these is physically relevant. The fact that the equations have other, spurious, solutions is a result of the fact that our equations are only a model of reality and some physical effects have been ignored. In particular, hyperbolic conservation laws do not include diffusive or viscous effects. Recall, for example, that the Euler equations result from the Navier-Stokes equations by ignoring fluid viscosity. Although viscous effects may be negligible throughout most of the flow, near discontinuities the effect is always strong. In fact, the full Navier-Stokes equations have smooth solutions for the simple flows we are considering, and the apparent discontinuities are in reality thin regions with very steep gradients. What we hope to model with the Euler equations is the limit of this smooth solution as the viscosity parameter approaches zero, which will in fact be one weak solution of the Euler equations.

Unfortunately there may be other weak solutions as well, and we must use our knowledge of what is being ignored in order to pick out the correct weak solution. This suggests the following general approach to defining a unique weak solution to a hyperbolic system of conservation laws: introduce a diffusive term into the equations to obtain an equation with a unique smooth solution, and then let the coefficient of this term go to zero. This "vanishing viscosity" method has some direct uses in the analysis of conservation laws, but is clearly not optimal since it requires studying a more complicated system of equations. This is precisely what we sought to avoid by introducing the inviscid equations in the first place. For this reason we would like to derive other conditions that can be imposed directly on weak solutions of the hyperbolic system to pick out the physically correct solution. For gas dynamics we can appeal to the second law of thermodynamics, which states that entropy is nondecreasing. In particular, as molecules of a gas pass through a shock their entropy should increase. It turns out that this condition is sufficient to recognize precisely those discontinuities that are physically correct and specify a unique solution.

For other systems of conservation laws it is frequently possible to derive similar conditions. These are generally called **entropy conditions** by analogy with gas dynamics.

Armed with the notion of weak solutions and an appropriate entropy condition, we can define mathematically a unique solution to the system of conservation laws that is the physically correct inviscid limit.

1.4   Numerical difficulties

When we attempt to calculate these solutions numerically, however, we face a new set of problems. We expect a finite difference discretization of the PDE to be inappropriate near discontinuities, where the PDE does not hold. Indeed, if we compute discontinuous solutions to conservation laws using standard methods developed under the assumption of smooth solutions, we typically obtain numerical results that are very poor.

As an example, Figure 1.4 shows some numerical results for the shock tube problem discussed earlier in this chapter. If we use Godunov's method, a first order accurate method described in Chapter 13, we obtain results that are very smeared in regions near the discontinuities. As we will see, this is because natural first order accurate numerical methods have a large amount of "numerical viscosity" that smoothes the solution in much the same way physical viscosity would, but to an extent that is unrealistic by several orders of magnitude. If we attempt to use a standard second order method, such as MacCormack's method (a version of the Lax-Wendroff method for conservation laws), we eliminate this numerical viscosity but now introduce dispersive effects that lead to large oscillations in the numerical solution, also seen in Figure 1.4. These results are very typical of what would be obtained with other standard first or second order methods.

Shock tracking. Since the PDEs continue to hold away from discontinuities, one possible approach is to combine a standard finite difference method in smooth regions with some explicit procedure for tracking the location of discontinuities. This is the numerical analogue of the mathematical approach in which the PDEs are supplemented by jump conditions across discontinuities. This approach is usually call "shock tracking". In one

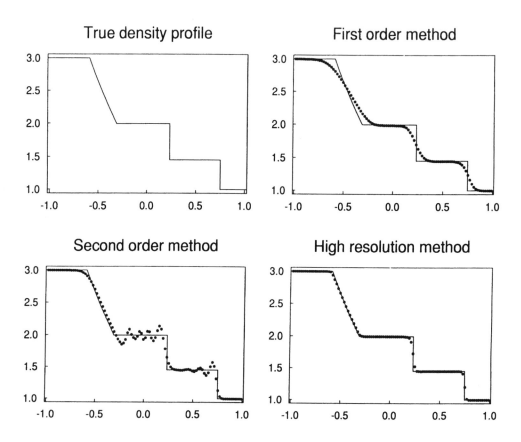

Figure 1.4. Solution of the shock tube problem at t = 0.5. The true density profile is shown along with the numerical solution computed with three different methods: Godunov's method (first order), MacCormack's method (second order), and a high resolution method.

space dimension it is often a viable approach. In more space dimensions the discontinuities typically lie along curves in 2D (recall Figure 1.2, for example) or surfaces in 3D and in realistic problems there may be many such surfaces that interact in complicated ways as time evolves. Although shock tracking is still possible, it becomes much more complicated and will not be pursued in these notes.

Shock capturing. Ideally we would like to have a numerical method that will produce sharp approximations to discontinuous solutions automatically, without explicit tracking and use of jump conditions. Methods that attempt to do this are called "shock capturing" methods. Over the past fifteen years a great deal of progress has been made in this direction, and today a variety of such methods are available. Some of the features we would like such a method to possess are:

- At least second order accuracy on smooth solutions, and also in smooth regions of a solution even when discontinuities are present elsewhere.

- Sharp resolution of discontinuities without excessive smearing.

- The absence of spurious oscillations in the computed solution.

- An appropriate form of consistency with the weak form of the conservation law, required if we hope to converge to weak solutions.

- Nonlinear stability bounds that, together with consistency, allow us to prove convergence as the grid is refined.

- A discrete form of the entropy condition, allowing us to conclude that the approximations in fact converge to the physically correct weak solution.

Methods with the accuracy and resolution properties indicated above are often referred to as **high resolution methods**. Our goal is to study the development and analysis of these methods. Figure 1.4 shows the results obtained on the shock tube problem with one of the high resolution methods described in Chapter 16.

Understanding these methods requires a good understanding of the mathematical theory of conservation laws, as well as some physical intuition about the behavior of solutions. Consequently, we will cover a considerable amount of material before we finally introduce these methods.

For linear hyperbolic systems **characteristics** play a major role. For nonlinear problems, the generalization of this theory which is most frequently used in developing numerical methods is the solution of a **Riemann problem**. This is simply the conservation law together with particular initial data consisting of two constant states separated by a single discontinuity,

$$u_0(x) = \begin{cases} u_l & x < 0, \\ u_r & x > 0. \end{cases} \tag{1.7}$$

For the Euler equations, this is simply the "shock tube problem" described previously. The solution of this problem has a relatively simple structure, and can be explicitly calculated in many cases. With a numerical method we compute a set of discrete values U_j^n, presumably approximations to $u(x_j, t_n)$ on some set of grid points $\{(x_j, t_n)\}$. A great deal of information about the local structure of the solution near (x_j, t_n) is obtained by solving a Riemann problem with data $u_l = U_j^n$ and $u_r = U_{j+1}^n$. Many numerical methods make use of these Riemann solutions, and so we study the Riemann problem and its solution in some detail in Chapters 6–9.

Obtaining the nonlinear stability results necessary to prove convergence is currently impossible for most methods in use for systems of equations. For the simpler case of a scalar equation, more analysis is possible. In particular, for many methods it is possible to show that the total variation of the solution is nonincreasing with time. This is enough to obtain some convergence results and also guarantees that spurious oscillations are not generated. Methods with this property are called **Total Variation Diminishing** Methods, or simply **TVD Methods**. Using this requirement as a design tool has lead to the development of many good numerical methods. Moreover, it is possible to generalize these scalar methods to systems of equations in a natural way using Riemann solutions. Even though the analysis and convergence proofs do not yet carry over, the resulting methods are often very successful in practice, having the "high resolution" properties that we desire.

Most practical problems are in two or three space dimensions and additional difficulties arise in this extension. There is currently some interest in the development of "fully multidimensional" methods that make use of the multidimensional structure of the solution locally, perhaps by some generalization of the Riemann problem. However, most of the methods currently in use are heavily based on one-dimensional methods, generalized by "dimensional splitting" or similar techniques. For this reason we will concentrate on one-dimensional methods, with some discussion at the end regarding generalizations. Of course solving multidimensional problems in complicated geometries introduces many other complications as well, such as grid generation and the proper treatment of boundary conditions. These topics are beyond the scope of these lectures. Several of the computational fluid dynamics books referred to below contain excellent discussions of these problems.

1.5   Some references

The next several chapters concern the mathematical theory of conservation laws. This material is self contained to a large extent, but the reader may wish to consult some of the excellent texts in this area for an alternative presentation or more details.

The elementary theory of linear hyperbolic equations is briefly reviewed in these notes, at least the aspects of particular importance in understanding the behavior of conservation laws. A more thorough treatment can be found in many standard books on PDEs, for example John[41]. This theory of hyperbolic equations with particular emphasis on the Euler and Navier-Stokes equations is also presented in Kreiss-Lorenz[42].

Some of the basic theory of nonlinear conservation laws is neatly presented in the monograph of Lax[45]. Majda[56] and Whitham[97] also present this material, with discussions of many applications including the Euler equations. The book of Courant-Friedrichs[11] deals almost entirely with gas dynamics and the Euler equations, but includes much of the general theory of conservation laws in this context and is very useful. The relatively new book of Smoller[77] includes more recent mathematical results not found in the other texts and is also an excellent reference in general.

Regarding numerical methods, many of the results and methods in the literature are not collected in any text, and so there will be frequent references to journal publications. For general background on numerical methods for PDEs, the new book of Strikwerda[82] is highly recommended. The classic book of Richtmyer-Morton[63] contains a good description of many of the mathematical techniques used to study numerical methods, particularly for linear equations. It also includes a large section on methods for nonlinear applications including fluid dynamics, but is out of date by now and does not discuss many of the methods we will study.

The recent book of Sod[79] describes some of the more modern methods and their mathematical analysis. Several other recent books on computational fluid dynamics are also good references, including Anderson-Tannehill-Pletcher[1], Boris-Oran[3], Fletcher[23], Hirsch[38], and Peyret-Taylor[61]. These books describe the fluid dynamics in more detail as well as the additional numerical problems arising in several space dimensions.

Finally, for an excellent collection of photographs illustrating a wide variety of interesting fluid dynamics, including shock waves, Van Dyke's *Album of Fluid Motion*[87] is highly recommended.

2 The Derivation of Conservation Laws

2.1 Integral and differential forms

To see how conservation laws arise from physical principles, we will begin by deriving the equation for conservation of mass in a one-dimensional gas dynamics problem, for example flow in a tube where properties of the gas such as density and velocity are assumed to be constant across each cross section of the tube. Let x represent the distance along the tube and let $\rho(x, t)$ be the density of the gas at point x and time t. This density is defined in such a way that the total mass of gas in any given section from x_1 to x_2, say, is given by the integral of the density:

$$\text{mass in } [x_1, x_2] \text{ at time } t = \int_{x_1}^{x_2} \rho(x, t) \, dx. \tag{2.1}$$

If we assume that the walls of the tube are impermeable and that mass is neither created nor destroyed, then the mass in this one section can change only because of gas flowing across the endpoints x_1 or x_2.

Now let $v(x, t)$ be the velocity of the gas at the point x at time t. Then the rate of flow, or **flux** of gas past this point is given by

$$\text{mass flux at } (x, t) = \rho(x, t)v(x, t). \tag{2.2}$$

By our comments above, the rate of change of mass in $[x_1, x_2]$ is given by the difference in fluxes at x_1 and x_2:

$$\frac{d}{dt} \int_{x_1}^{x_2} \rho(x, t) \, dx = \rho(x_1, t)v(x_1, t) - \rho(x_2, t)v(x_2, t). \tag{2.3}$$

This is one **integral form** of the conservation law. Another form is obtained by integrating this in time from t_1 to t_2, giving an expression for the mass in $[x_1, x_2]$ at time $t_2 > t_1$ in terms of the mass at time t_1 and the total (integrated) flux at each boundary during

14

this time period:

$$\int_{x_1}^{x_2} \rho(x, t_2) \, dx = \int_{x_1}^{x_2} \rho(x, t_1) \, dx \tag{2.4}$$
$$+ \int_{t_1}^{t_2} \rho(x_1, t) v(x_1, t) \, dt - \int_{t_1}^{t_2} \rho(x_2, t) v(x_2, t) \, dt.$$

To derive the differential form of the conservation law, we must now assume that $\rho(x, t)$ and $v(x, t)$ are differentiable functions. Then using

$$\rho(x, t_2) - \rho(x, t_1) = \int_{t_1}^{t_2} \frac{\partial}{\partial t} \rho(x, t) \, dt \tag{2.5}$$

and

$$\rho(x_2, t) v(x_2, t) - \rho(x_1, t) v(x_1, t) = \int_{x_1}^{x_2} \frac{\partial}{\partial x} (\rho(x, t) v(x, t)) \, dx \tag{2.6}$$

in (2.4) gives

$$\int_{t_1}^{t_2} \int_{x_1}^{x_2} \left\{ \frac{\partial}{\partial t} \rho(x, t) + \frac{\partial}{\partial x} (\rho(x, t) v(x, t)) \right\} \, dx \, dt = 0. \tag{2.7}$$

Since this must hold for any section $[x_1, x_2]$ and over any time interval $[t_1, t_2]$, we conclude that in fact the integrand in (2.7) must be identically zero, i.e.,

$$\rho_t + (\rho v)_x = 0 \qquad \text{conservation of mass.} \tag{2.8}$$

This is the desired **differential form** of the conservation law for the conservation of mass.

The conservation law (2.8) can be solved in isolation only if the velocity $v(x, t)$ is known a *priori* or is known as a function of $\rho(x, t)$. If it is, then ρv is a function of ρ alone, say $\rho v = f(\rho)$, and the conservation of mass equation (2.8) becomes a scalar conservation law for ρ,

$$\rho_t + f(\rho)_x = 0. \tag{2.9}$$

More typically the equation (2.8) must be solved in conjunction with equations for the conservation of momentum and energy. These equations will be derived and discussed in more detail in Chapter 5. For now we simply state them for the case of the **Euler equations** of gas dynamics:

$$(\rho v)_t + (\rho v^2 + p)_x = 0 \qquad \text{conservation of momentum} \tag{2.10}$$

$$E_t + (v(E + p))_x = 0 \qquad \text{conservation of energy} \tag{2.11}$$

Note that these equatons involve another quantity, the pressure p, which must be specified as a given function of ρ, ρv, and E in order that the fluxes are well defined functions of the conserved quantities alone. This additional equation is called the **equation of state** and depends on physical properties of the gas under study.

If we introduce the vector $u \in \mathbb{R}^3$ as

$$u(x,t) = \begin{bmatrix} \rho(x,t) \\ \rho(x,t)v(x,t) \\ E(x,t) \end{bmatrix} \equiv \begin{bmatrix} u_1 \\ u_2 \\ u_3 \end{bmatrix}, \tag{2.12}$$

then the system of equations (2.8), (2.10), (2.11) can be written simply as

$$u_t + f(u)_x = 0 \tag{2.13}$$

where

$$f(u) = \begin{bmatrix} \rho v \\ \rho v^2 + p \\ v(E+p) \end{bmatrix} = \begin{bmatrix} u_2 \\ u_2^2/u_1 + p(u) \\ u_2(u_3 + p(u))/u_1 \end{bmatrix}. \tag{2.14}$$

Again, the form (2.13) is the differential form of the conservation laws, which holds in the usual sense only where u is smooth. More generally, the integral form for a system of m equations says that

$$\frac{d}{dt} \int_{x_1}^{x_2} u(x,t)\, dx = f(u(x_1,t)) - f(u(x_2,t)) \tag{2.15}$$

for all x_1, x_2, t. Equivalently, integrating from t_1 to t_2 gives

$$\begin{aligned}
\int_{x_1}^{x_2} u(x,t_2)\, dx &= \int_{x_1}^{x_2} u(x,t_1)\, dx \\
&\quad + \int_{t_1}^{t_2} f(u(x_1,t))\, dt - \int_{t_1}^{t_2} f(u(x_2,t))\, dt
\end{aligned} \tag{2.16}$$

for all x_1, x_2, t_1, and t_2. These integral forms of the conservation law will be fundamental in later analysis.

2.2   Scalar equations

Before tackling the complications of coupled systems of equations, we will first study the case of a scalar equation, where $m = 1$. An example is the conservation of mass equation (2.8) in the case where v is a known function of $\rho(x,t)$. This does not happen in gas dynamics, but can occur in other problems where the same conservation law holds. One example is a simple model of traffic flow along a highway. Here ρ is the density of vehicles and the velocity at which people drive is assumed to depend only on the local density. This example is explored in detail in Chapter 4.

Another possibility is that the velocity $v(x,t)$ is given a priori, completely independent of the unknown ρ. Suppose, for example, that ρ represents the density of some chemical in flowing water, a pollutant in a stream, for example. Since the total quantity of this chemical is conserved, the same derivation as above again yields (2.8). Now it is reasonable to consider a case where v, the fluid velocity, is known from other sources and can be assumed given. Changes in concentration of the chemical will have little or no effect on the fluid dynamics and so there will be no coupling in which v depends on ρ.

Strictly speaking, this conservation law is not of the form (1.1), since the flux function now depends explicitly on x and t as well as on ρ,

$$f(\rho, x, t) = \rho v(x, t). \tag{2.17}$$

However, the conservation law is still hyperbolic and this variable coefficient linear equation is convenient to study as an introductory example. Moreover, if the velocity is constant, $u(x, t) \equiv a$, then $f(\rho) = a\rho$ and (2.8) reduces to

$$\rho_t + a\rho_x = 0. \tag{2.18}$$

This equation is known as the **linear advection equation** or sometimes the **one-way wave equation**. If this equation is solved for $t \geq 0$ with initial data

$$\rho(x, 0) = \rho_0(x) \qquad -\infty < x < \infty \tag{2.19}$$

then it is easy to check (assuming ρ_0 is differentiable) that the solution is simply

$$\rho(x, t) = \rho_0(x - at). \tag{2.20}$$

Note that the initial profile simply moves downstream with velocity a, its shape unchanged. It is reasonable to say that the solution is given by (2.20) even if the initial data $\rho_0(x)$ is not smooth, as will be justified in Chapter 3.

2.3 Diffusion

In the example just given of a concentration profile moving downstream, the result that this profile remains unchanged may seem physically unreasonable. We expect mixing and diffusion to occur as well as advection. Mixing caused by turbulence in the fluid is clearly ignored in this simple one-dimensional model, but molecular diffusion should occur even in a stream moving at constant velocity, since by "velocity" we really mean the macroscopic average velocity of the molecules. Individual molecules travel with random directions and speeds distributed about this average. In particular, the molecules of our chemical undergo this random motion and hence there is a net motion of these molecules away from regions of high concentration and towards regions of lower concentration. Since the chemical is still conserved, we can incorporate this effect into the conservation law by modifying the flux function appropriately.

Consider the flux of the chemical past some point x in the tube or stream. In addition to the advective flux $a\rho$ there is also a net flux due to diffusion whenever the concentration profile is not flat at the point x. This flux is determined by "Fourier's Law of heat conduction" (heat diffuses in much the same way as the chemical concentration), which says that the diffusive flux is simply proportional to the gradient of concentration:

$$\text{diffusive flux} = -D\rho_x. \tag{2.21}$$

The diffusion coefficient $D > 0$ depends on the variance of the random component of the particle speeds. The minus sign in (2.21) is needed since the net flux is *away* from higher concentrations.

Combining this flux with the advective flux $a\rho$ gives the flux function

$$f(\rho, \rho_x) = a\rho - D\rho_x \tag{2.22}$$

and the conservation law becomes

$$\rho_t + (a\rho - D\rho_x)_x = 0 \tag{2.23}$$

or, assuming D is constant,

$$\rho_t + a\rho_x = D\rho_{xx}. \tag{2.24}$$

Equation (2.24) is called the **advection-diffusion equation** (or sometimes the convection-diffusion equation).

Note that the flux function $f(\rho, \rho_x)$ now depends on ρ_x as well as ρ. Equation (2.23), while still a conservation law, is not of the form (1.1). In this case, the dependence of f on the gradient completely changes the character of the equation and solution. The advection-diffusion equation (2.24) is a parabolic equation while (2.18) is hyperbolic. One major difference is that (2.24) always has smooth solutions for $t > 0$ even if the initial data $\rho_0(x)$ is discontinuous. This is analogous to the relation between the Euler equations and Navier-Stokes equations described in Chapter 1. We can view (2.18) as an approximation to (2.24) valid for D very small, but we may need to consider the effect of D in order to properly interpret discontinuous solutions to (2.18).

3 Scalar Conservation Laws

We begin our study of conservation laws by considering the scalar case. Many of the difficulties encountered with systems of equations are already encountered here, and a good understanding of the scalar equation is required before proceeding.

3.1 The linear advection equation

We first consider the linear advection equation, derived in Chapter 2, which we now write as

$$u_t + au_x = 0. \tag{3.1}$$

The Cauchy problem is defined by this equation on the domain $-\infty < x < \infty$, $t \geq 0$ together with initial conditions

$$u(x, 0) = u_0(x). \tag{3.2}$$

As noted previously, the solution is simply

$$u(x, t) = u_0(x - at) \tag{3.3}$$

for $t \geq 0$. As time evolves, the initial data simply propagates unchanged to the right (if $a > 0$) or left (if $a < 0$) with velocity a. The solution $u(x, t)$ is constant along each ray $x - at = x_0$, which are known as the **characteristics** of the equation. (See Fig. 3.1 for the case $a > 0$.)

Note that the characteristics are curves in the x-t plane satisfying the ordinary differential equations $x'(t) = a$, $x(0) = x_0$. If we differentiate $u(x, t)$ along one of these curves to find the rate of change of u along the characteristic, we find that

$$\begin{aligned}
\frac{d}{dt}u(x(t), t) &= \frac{\partial}{\partial t}u(x(t), t) + \frac{\partial}{\partial x}u(x(t), t)\, x'(t) \\
&= u_t + au_x \\
&= 0,
\end{aligned} \tag{3.4}$$

confirming that u is constant along these characteristics.

More generally, we might consider a variable coefficient advection equation of the form

$$u_t + (a(x)u)_x = 0, \tag{3.5}$$

where $a(x)$ is a smooth function. Recalling the derivation of the advection equation in Chapter 2, this models the evolution of a chemical concentration $u(x,t)$ in a stream with variable velocity $a(x)$.

We can rewrite (3.5) as

$$u_t + a(x)u_x = -a'(x)u \tag{3.6}$$

or

$$\left(\frac{\partial}{\partial t} + a(x)\frac{\partial}{\partial x} \right) u(x,t) = -a'(x)u(x,t). \tag{3.7}$$

It follows that the evolution of u along any curve $x(t)$ satisfying

$$\begin{aligned} x'(t) &= a(x(t)) \\ x(0) &= x_0 \end{aligned} \tag{3.8}$$

satisfies a simple ordinary differential equation (ODE):

$$\frac{d}{dt}u(x(t),t) = -a'(x(t))\, u(x(t),t). \tag{3.9}$$

The curves determined by (3.8) are again called characteristics. In this case the solution $u(x,t)$ is not constant along these curves, but can be easily determined by solving two sets of ODEs.

It can be shown that if $u_0(x)$ is a smooth function, say $u_0 \in C^k(-\infty,\infty)$, then the solution $u(x,t)$ is equally smooth in space and time, $u \in C^k((-\infty,\infty) \times (0,\infty))$.

3.1.1 Domain of dependence

Note that solutions to the linear advection equations (3.1) and (3.5) have the following property: the solution $u(x,t)$ at any point $(\bar{x},\bar{t})$ depends on the initial data u_0 only at a *single* point, namely the point $\bar{x}_0$ such that $(\bar{x},\bar{t})$ lies on the characteristic through $\bar{x}_0$. We could change the initial data at any points other than $\bar{x}_0$ without affecting the solution $u(\bar{x},\bar{t})$. The set $\mathcal{D}(\bar{x},\bar{t}) = \{\bar{x}_0\}$ is called the **domain of dependence** of the point $(\bar{x},\bar{t})$. Here this domain consists of a single point. For a system of equations this domain is typically an interval, but a fundamental fact about hyperbolic equations is that it is always a *bounded* interval. The solution at $(\bar{x},\bar{t})$ is determined by the initial data within some finite distance of the point $\bar{x}$. The size of this set usually increases with $\bar{t}$, but we have a bound of the form $\mathcal{D} \subset \{x : |x - \bar{x}| \le a_{max}\bar{t}\}$ for some value a_{max}. Conversely, initial data at any given point x_0 can influence the solution only within some

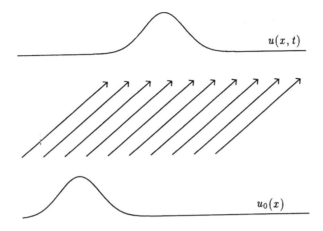

Figure 3.1. Characteristics and solution for the advection equation.

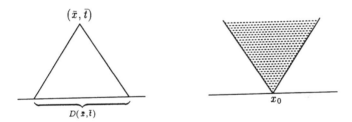

Figure 3.2. Domain of dependence and range of influence.

cone $\{x : |x - x_0| \leq a_{\max}t\}$ of the x-t plane. This region is called the **range of influence** of the point x_0. See Figure 3.2 for an illustration. We summarize this by saying that hyperbolic equations have **finite propagation speed**; information can travel with speed at most $a_{\max}$. This has important consequences in developing numerical methods.

3.1.2   Nonsmooth data

In the manipulations performed above, we have assumed differentiability of $u(x, t)$. However, from our observation that the solution along a characteristic curve depends only on the one value $u_0(x_0)$, it is clear that spatial smoothness is not required for this construction of $u(x, t)$ from $u_0(x)$. We can thus define a "solution" to the PDE even if $u_0(x)$ is not a smooth function. Note that if $u_0(x)$ has a singularity at some point x_0 (a discontinuity in u_0 or some derivative), then the resulting $u(x, t)$ will have a singularity of the same order along the characteristic curve through x_0, but will remain smooth along characteristics through smooth portions of the data. This is a fundamental property of *linear* hyperbolic equations: singularities propagate only along characteristics.

If u_0 is nondifferentiable at some point then $u(x, t)$ is no longer a classical solution of the differential equation everywhere. However, this function *does* satisfy the integral form of the conservation law, which continues to make sense for nonsmooth u. Recall that the integral form is more fundamental physically than the differential equation, which was derived from the integral form under the additional assumption of smoothness. It thus makes perfect sense to accept this generalized solution as solving the conservation law.

EXERCISE 3.1.   *Let $f(u) = au$, with a constant, and let $u_0(x)$ be any integrable function. Verify that the function $u(x, t) = u_0(x - at)$ satisfies the integral form (2.16) for any x_1, x_2, t_1 and t_2.*

Other approaches can also be taken to defining this generalized solution, which extend better to the study of nonlinear equations where we can no longer simply integrate along characteristics.

One possibility is to approximate the nonsmooth data $u_0(x)$ by a sequence of smooth functions $u_0^\epsilon(x)$, with

$$\| u_0 - u_0^\epsilon \|_1 < \epsilon$$

as $\epsilon \to 0$. Here $\| \cdot \|_1$ is the 1-norm, defined by

$$\| v \|_1 = \int_{-\infty}^{\infty} |v(x)| \, dx. \tag{3.10}$$

For the linear equation we know that the PDE together with the smooth data u_0^ϵ has a smooth classical solution $u^\epsilon(x, t)$ for all $t \geq 0$. We can now define the generalized solution $u(x, t)$ by taking the limit of $u^\epsilon(x, t)$ as $\epsilon \to 0$. For example, the constant coefficient problem (3.1) has classical smooth solutions

$$u^\epsilon(x, t) = u_0^\epsilon(x - at)$$

and clearly at each time t the 1-norm limit exists and satisfies

$$u(x, t) = \lim_{\epsilon \to 0} u_0^\epsilon(x - at) = u_0(x - at)$$

as expected.

Unfortunately, this approach of smoothing the initial data will not work for nonlinear problems. As we will see, solutions to the nonlinear problem can develop singularities even if the initial data is smooth, and so there is no guarantee that classical solutions with data $u_0^\epsilon(x)$ will exist.

A better approach, which does generalize to nonlinear equations, is to leave the initial data alone but modify the PDE by adding a small diffusive term. Recall from Chapter 2 that the conservation law (3.1) should be considered as an approximation to the advection-diffusion equation

$$u_t + au_x = \epsilon u_{xx} \tag{3.11}$$

for ϵ very small. If we now let $u^\epsilon(x,t)$ denote the solution of (3.11) with data $u_0(x)$, then $u^\epsilon \in C^\infty((-\infty,\infty) \times (0,\infty))$ even if $u_0(x)$ is not smooth since (3.11) is a parabolic equation. We can again take the limit of $u^\epsilon(x,t)$ as $\epsilon \to 0$, and will obtain the same generalized solution $u(x,t)$ as before.

Note that the equation (3.11) simplifies if we make a change of variables to follow the characteristics, setting

$$v^\epsilon(x,t) = u^\epsilon(x+at,t).$$

Then it is easy to verify that v^ϵ satsifies the **heat equation**

$$v_t^\epsilon(x,t) = \epsilon v_{xx}^\epsilon(x,t). \tag{3.12}$$

Using the well-known solution to the heat equation to solve for $v(x,t)$, we have $u^\epsilon(x,t) = v^\epsilon(x-at,t)$ and so can explicitly calculate the "vanishing viscosity" solution in this case.

EXERCISE 3.2. *Show that the vanishing viscosity solution* $\lim_{\epsilon \to 0} u^\epsilon(x,t)$ *is equal to* $u_0(x-at)$.

3.2 Burgers' equation

Now consider the nonlinear scalar equation

$$u_t + f(u)_x = 0 \tag{3.13}$$

where $f(u)$ is a nonlinear function of u. We will assume for the most part that $f(u)$ is a convex function, $f''(u) > 0$ for all u (or, equally well, f is concave with $f''(u) < 0 \; \forall u$). The convexity assumption corresponds to a "genuine nonlinearity" assumption for systems of equations that holds in many important cases, such as the Euler equations. The nonconvex case is also important in some applications (e.g. oil reservoir simulation) but is more complicated mathematically. One nonconvex example, the Buckley-Leverett equation, is discussed in the next chapter.

By far the most famous model problem in this field is **Burgers' equation**, in which $f(u) = \frac{1}{2}u^2$, so (3.13) becomes

$$u_t + uu_x = 0. \tag{3.14}$$

Actually this should be called the "inviscid Burgers' equation", since the equation studied by Burgers[5] also includes a viscous term:

$$u_t + uu_x = \epsilon u_{xx}. \tag{3.15}$$

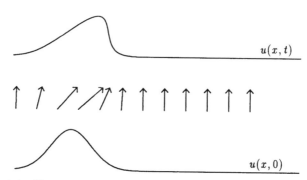

Figure 3.3. Characteristics and solution for Burgers' equation (small t).

This is about the simplest model that includes the nonlinear and viscous effects of fluid dynamics.

Around 1950, Hopf, and independently Cole, showed that the *exact* solution of the nonlinear equation (3.15) could be found using what is now called the **Cole-Hopf trans-formation**. This reduces (3.15) to a linear heat equation. See Chapter 4 of Whitham[97] for details.

Consider the inviscid equation (3.14) with smooth initial data. For small time, a solution can be constructed by following characteristics. Note that (3.14) looks like an advection equation, but with the advection velocity u equal to the value of the advected quantity. The characteristics satisfy

$$x'(t) = u(x(t), t) \tag{3.16}$$

and along each characteristic u is constant, since

$$
\begin{aligned}
\frac{d}{dt} u(x(t), t) &= \frac{\partial}{\partial t} u(x(t), t) + \frac{\partial}{\partial x} u(x(t), t)\, x'(t) \\
&= u_t + u u_x \\
&= 0.
\end{aligned}
\tag{3.17}
$$

Moreover, since u is constant on each characteristic, the slope $x'(t)$ is constant by (3.16) and so the characteristics are straight lines, determined by the initial data (see Fig. 3.3).

If the initial data is smooth, then this can be used to determine the solution $u(x, t)$ for small enough t that characteristics do not cross: For each (x, t) we can solve the equation

$$x = \xi + u(\xi, 0)t \tag{3.18}$$

for ξ and then

$$u(x, t) = u(\xi, 0). \tag{3.19}$$

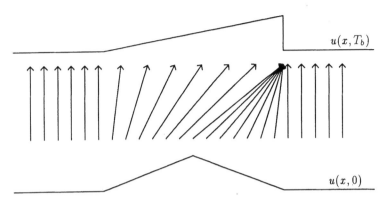

Figure 3.4. Shock formation in Burgers' equation.

3.3   Shock formation

For larger t the equation (3.18) may not have a unique solution. This happens when the characteristics cross, as will eventually happen if $u_x(x,0)$ is negative at any point. At the time T_b where the characteristics first cross, the function $u(x,t)$ has an infinite slope — the wave "breaks" and a shock forms. Beyond this point there is no classical solution of the PDE, and the weak solution we wish to determine becomes discontinuous.

Figure 3.4 shows an extreme example where the initial data is piecewise linear and many characteristics come together at once. More generally an infinite slope in the solution may appear first at just one point x, corresponding via (3.18) to the point ξ where the slope of the initial data is most negative. At this point the wave is said to "break", by analogy with waves on a beach.

EXERCISE 3.3. *Show that if we solve (3.14) with smooth initial data $u_0(x)$ for which $u_0'(x)$ is somewhere negative, then the wave will break at time*

$$T_b = \frac{-1}{\min u_0'(x)}. \tag{3.20}$$

Generalize this to arbitrary convex scalar equations.

For times $t > T_b$ some of the characteristics have crossed and so there are points x where there are three characteristics leading back to $t = 0$. One can view the "solution" u at such a time as a triple-valued function (see Fig. 3.5).

This sort of solution makes sense in some contexts, for example a breaking wave on a sloping beach can be modeled by the shallow water equations and, for a while at least, does behave as seen in Fig. 3.5, with fluid depth a triple-valued function.

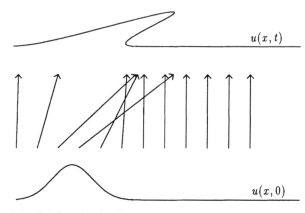

Figure 3.5. Triple-valued solution to Burgers' equation at time $t > T_b$.

Figure 3.6. Solution to the viscous Burgers' equation at time T_b for the data shown in Figure 3.4.

However, in most physical situations this does not make sense. For example, the density of a gas cannot possibly be triple valued at a point. What happens instead at time T_b?

We can determine the correct physical behavior by adopting the vanishing viscosity approach. The equation (3.14) is a model of (3.15) valid only for small ϵ and smooth u. When it breaks down, we must return to (3.15). If the initial data is smooth and ϵ very small, then before the wave begins to break the ϵu_{xx} term is negligible compared to the other terms and the solutions to both PDEs look nearly identical. Figure 3.3, for example, would be essentially unchanged if we solved (3.15) with small ϵ rather than (3.14). However, as the wave begins to break, the second derivative term u_{xx} grows much faster than u_x, and at some point the ϵu_{xx} term is comparable to the other terms and begins to play a role. This term keeps the solution smooth for all time, preventing the breakdown of solutions that occurs for the hyperbolic problem.

For very small values of ϵ, the discontinuous solution at T_b shown in Figure 3.4 would

Figure 3.7. Solution to the viscous Burgers' equation for two different values of ϵ.

be replaced by a smooth continuous function as in Figure 3.6. As $\epsilon \to 0$ this becomes sharper and approaches the discontinuous solution of Figure 3.4.

For times $t > T_b$, such as was shown in Figure 3.5, the viscous solution for $\epsilon > 0$ would continue to be smooth and single valued, with a shape similar to that shown in Figure 3.6. The behavior as $\epsilon \to 0$ is indicated in Figure 3.7.

It is this vanishing viscosity solution that we hope to capture by solving the inviscid equation.

3.4   Weak solutions

A natural way to define a generalized solution of the inviscid equation that does not require differentiability is to go back to the integral form of the conservation law, and say that $u(x, t)$ is a generalized solution if (2.7) is satisfied for all x_1, x_2, t_1, t_2.

There is another approach that results in a different integral formulation that is often more convenient to work with. This is a mathematical technique that can be applied more generally to rewrite a differential equation in a form where less smoothness is required to define a "solution". The basic idea is to take the PDE, multiply by a smooth "test function", integrate one or more times over some domain, and then use integration by parts to move derivatives off the function u and onto the smooth test function. The result is an equation involving fewer derivatives on u, and hence requiring less smoothness.

In our case we will use test functions $\phi \in C_0^1(\mathbb{R} \times \mathbb{R})$. Here C_0^1 is the space of function that are continuously differentiable with "compact support". The latter requirement means that $\phi(x, t)$ is identically zero outside of some bounded set, and so the support of the function lies in a compact set.

If we multiply $u_t + f_x = 0$ by $\phi(x, t)$ and then integrate over space and time, we obtain

$$\int_0^\infty \int_{-\infty}^{+\infty} [\phi u_t + \phi f(u)_x] \, dx \, dt = 0. \tag{3.21}$$

Now integrate by parts, yielding

$$\int_0^\infty \int_{-\infty}^{+\infty} [\phi_t u + \phi_x f(u)] \, dx \, dt = -\int_{-\infty}^\infty \phi(x, 0) u(x, 0) \, dx. \tag{3.22}$$

Note that nearly all the boundary terms which normally arise through integration by parts drop out due to the requirement that ϕ have compact support, and hence vanishes at infinity. The remaining boundary term brings in the initial conditions of the PDE, which must still play a role in our weak formulation.

DEFINITION 3.1. *The function $u(x,t)$ is called a weak solution of the conservation law if (3.22) holds for all functions $\phi \in C_0^1(\mathbb{R} \times \mathbb{R}^+)$.*

The advantage of this formulation over the original integral form (2.16) is that the integration in (3.22) is over a fixed domain, all of $\mathbb{R} \times \mathbb{R}^+$. The integral form (2.16) is over an arbitrary rectangle, and to check that $u(x,t)$ is a solution we must verify that this holds for all choices of x_1, x_2, t_1 and t_2. Of course, our new form (3.22) has a similar feature, we must check that it holds for all $\phi \in C_0^1$, but this turns out to be an easier task.

Mathematically the two integral forms are equivalent and we should rightly expect a more direct connection between the two that does not involve the differential equation. This can be achieved by considering special test functions $\phi(x,t)$ with the property that

$$\phi(x,t) = \begin{cases} 1 & \text{for } (x,t) \in [x_1, x_2] \times [t_1, t_2] \\ 0 & \text{for } (x,t) \notin [x_1 - \epsilon, x_2 + \epsilon] \times [t_1 - \epsilon, t_2 + \epsilon] \end{cases} \tag{3.23}$$

and with ϕ smooth in the intermediate strip of width ϵ. Then $\phi_x = \phi_t = 0$ except in this strip and so the integral (3.22) reduces to an integral over this strip. As $\epsilon \to 0$, ϕ_x and ϕ_t approach delta functions that sample u or $f(u)$ along the boundaries of the rectangle $[x_1, x_2] \times [t_1, t_2]$, so that (3.22) approaches the integral form (2.16). By making this rigorous, we can show that any weak solution satisfies the original integral conservation law.

The vanishing viscosity generalized solution we defined above is a weak solution in the sense of (3.22), and so this definition includes the solution we are looking for. Unfortunately, weak solutions are often not unique, and so an additional problem is often to identify *which* weak solution is the physically correct vanishing viscosity solution. Again, one would like to avoid working with the viscous equation directly, but it turns out that there are other conditions one can impose on weak solutions that are easier to check and will also pick out the correct solution. As noted in Chapter 1, these are usually called *entropy conditions* by analogy with the gas dynamics case. The vanishing viscosity solution is also called the **entropy solution** because of this.

3.5   The Riemann Problem

The conservation law together with piecewise constant data having a single discontinuity is known as the Riemann problem. As an example, consider Burgers' equation $u_t + uu_x = 0$

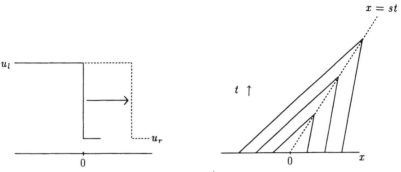

Figure 3.8. Shock wave.

with piecewise constant initial data

$$u(x,0) = \begin{cases} u_l & x < 0 \\ u_r & x > 0. \end{cases} \qquad (3.24)$$

The form of the solution depends on the relation between u_l and u_r.

Case I. $u_l > u_r$.

In this case there is a unique weak solution,

$$u(x,t) = \begin{cases} u_l & x < st \\ u_r & x > st \end{cases} \qquad (3.25)$$

where

$$s = (u_l + u_r)/2 \qquad (3.26)$$

is the **shock speed**, the speed at which the discontinuity travels. A general expression for the shock speed will be derived below. Note that characteristics in each of the regions where u is constant go *into* the shock (see Fig. 3.8) as time advances.

EXERCISE 3.4. *Verify that (3.25) is a weak solution to Burgers' equation by showing that (3.22) is satisfied for all $\phi \in C_0^1$.*

EXERCISE 3.5. *Show that the viscous equation (3.15) has a travelling wave solution of the form $u^\epsilon(x,t) = w(x - st)$ by deriving an ODE for w and verifying that this ODE has solutions of the form*

$$w(y) = u_r + \frac{1}{2}(u_l - u_r)\left[1 - \tanh((u_l - u_r)y/4\epsilon)\right] \qquad (3.27)$$

with s again given by (3.26). Note that $w(y) \to u_l$ as $y \to -\infty$ and $w(y) \to u_r$ as $y \to +\infty$. Sketch this solution and indicate how it varies as $\epsilon \to 0$.

The smooth solution $u^\epsilon(x,t)$ found in Exercise 3.5 converges to the shock solution (3.25) as $\epsilon \to 0$, showing that our shock solution is the desired vanishing viscosity solution. The shape of $u^\epsilon(x,t)$ is often called the "viscous profile" for the shock wave.

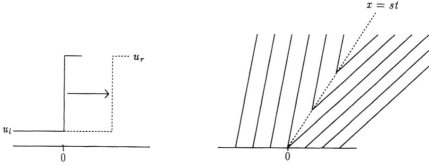

Figure 3.9. Entropy-violating shock.

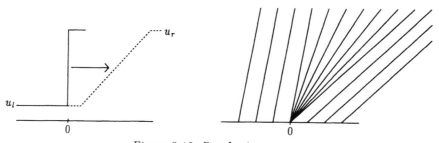

Figure 3.10. Rarefaction wave.

Case II. $u_l < u_r$.

In this case there are infinitely many weak solutions. One of these is again (3.25), (3.26) in which the discontinuity propagates with speed s. Note that characteristics now go *out* of the shock (Fig. 3.9) and that this solution is not stable to perturbations. If the data is smeared out slightly, or if a small amount of viscosity is added to the equation, the solution changes completely.

Another weak solution is the **rarefaction wave**

$$u(x,t) = \begin{cases} u_l & x < u_l t \\ x/t & u_l t \le x \le u_r t \\ u_r & x > u_r t \end{cases} \tag{3.28}$$

This solution is stable to perturbations and is in fact the vanishing viscosity generalized solution (Fig. 3.10).

EXERCISE 3.6. *There are infinitely many other weak solutions to (3.14) when $u_l < u_r$. Show, for example, that*

$$u(x,t) = \begin{cases} u_l & x < s_m t \\ u_m & s_m t \le x \le u_m t \\ x/t & u_m t \le x \le u_r t \\ u_r & x > u_r t \end{cases}$$

is a weak solution for any u_m with $u_l \leq u_m \leq u_r$ and $s_m = (u_l + u_m)/2$. Sketch the characteristics for this solution. Find a class of weak solutions with three discontinuities.

EXERCISE 3.7. *Show that for a general convex scalar problem (3.13) with data (3.24) and $u_l < u_r$, the rarefaction wave solution is given by*

$$u(x, t) = \begin{cases} u_l & x < f'(u_l)t \\ v(x/t) & f'(u_l)t \leq x \leq f'(u_r)t \\ u_r & x > f'(u_r)t \end{cases} \qquad (3.29)$$

where $v(\xi)$ is the solution to $f'(v(\xi)) = \xi$.

3.6 Shock speed

The propagating shock solution (3.25) is a weak solution to Burgers' equation only if the speed of propagation is given by (3.26). The same discontinuity propagating at a different speed would not be a weak solution.

The speed of propagation can be determined by conservation. If M is large compared to st then by (2.15), $\int_{-M}^{M} u(x, t)\, dx$ must increase at the rate

$$\frac{d}{dt} \int_{-M}^{M} u(x, t)\, dx = f(u_l) - f(u_r) \qquad (3.30)$$

$$= \frac{1}{2}(u_l + u_r)(u_l - u_r)$$

for Burgers' equation. On the other hand, the solution (3.25) clearly has

$$\int_{-M}^{M} u(x, t)\, dx = (M + st)u_l + (M - st)u_r \qquad (3.31)$$

so that

$$\frac{d}{dt} \int_{-M}^{M} u(x, t)\, dx = s(u_l - u_r). \qquad (3.32)$$

Comparing (3.30) and (3.32) gives (3.26).

More generally, for arbitrary flux function $f(u)$ this same argument gives the following relation between the shock speed s and the states u_l and u_r, called the **Rankine-Hugoniot jump condition**:

$$f(u_l) - f(u_r) = s(u_l - u_r). \qquad (3.33)$$

For scalar problems this gives simply

$$s = \frac{f(u_l) - f(u_r)}{u_l - u_r} = \frac{[f]}{[u]} \qquad (3.34)$$

where $[\cdot]$ indicates the jump in some quantity across the discontinuity. Note that any jump is allowed, provided the speed is related via (3.34).

For systems of equations, $u_l - u_r$ and $f(u_r) - f(u_l)$ are both vectors while s is still a scalar. Now we cannot always solve for s to make (3.33) hold. Instead, only certain jumps $u_l - u_r$ are allowed, namely those for which the vectors $f(u_l) - f(u_r)$ and $u_l - u_r$ are linearly dependent.

EXAMPLE 3.1.   For a linear system with $f(u) = Au$, (3.33) gives

$$A(u_l - u_r) = s(u_l - u_r),\tag{3.35}$$

i.e., $u_l - u_r$ must be an eigenvector of the matrix A and s is the associated eigenvalue. For a linear system, these eigenvalues are the characteristic speeds of the system. Thus discontinuities can propagate only along characteristics, a fact that we have already seen for the scalar case.

So far we have considered only piecewise constant initial data and shock solutions consisting of a single discontinuity propagating at constant speed. More typically, solutions have both smooth regions, where the PDEs are satisfied in the classical sense, and propagating discontinuities whose strength and speed vary as they interact with the smooth flow or collide with other shocks.

The Rankine-Hugoniot (R-H) conditions (3.33) hold more generally across any propagating shock, where now u_l and u_r denote the values immediately to the left and right of the discontinuity and s is the corresponding instantaneous speed, which varies along with u_l and u_r.

EXAMPLE 3.2.   As an example, the following "N wave" is a solution to Burgers' equation:

$$u(x,t) = \begin{cases} x/t & -\sqrt{t} < x < \sqrt{t} \\ 0 & \text{otherwise} \end{cases}\tag{3.36}$$

This solution has two shocks propagating with speeds $\pm\frac{1}{2\sqrt{t}}$. The right-going shock has left and right states $u_l = \sqrt{t}/t = 1/\sqrt{t}$, $u_r = 0$ and so the R-H condition is satisfied, and similarly for the left-going shock. See Figure 3.11.

To verify that the R-H condition must be instantaneously satisfied when u_l and u_r vary, we apply the same conservation argument as before but now to a small rectangle as shown in Figure 3.12, with $x_2 = x_1 + \Delta x$ and $t_2 = t_1 + \Delta t$. Assuming that u is smoothly varying on each side of the shock, and that the shock speed $s(t)$ is consequently also smoothly varying, we have the following relation between Δx and Δt:

$$\Delta x = s(t_1)\Delta t + O(\Delta t^2).\tag{3.37}$$

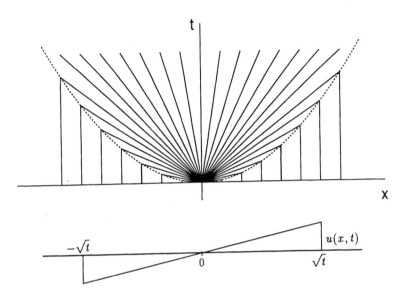

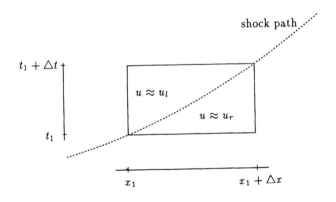

Figure 3.11. N wave solution to Burgers' equation.

Figure 3.12. Region of integration for shock speed calculation.

From the integral form of the conservation law, we have

$$\int_{x_1}^{x_1+\Delta x} u(x, t_1 + \Delta t)\, dx = \int_{x_1}^{x_1+\Delta x} u(x, t_1)\, dx \tag{3.38}$$

$$+ \int_{t_1}^{t_1+\Delta t} f(u(x_1, t))\, dt - \int_{t_1}^{t_1+\Delta t} f(u(x_1 + \Delta x, t))\, dt.$$

In the triangular portion of the infinitesimal rectangle that lies to the left of the shock, $u(x, t) = u_l(t_1) + O(\Delta t)$, while in the complementary triangle, $u(x, t) = u_r(t_1) + O(\Delta t)$. Using this in (3.38) gives

$$\Delta x\, u_l = \Delta x\, u_r + \Delta t\, f(u_l) - \Delta t\, f(u_r) + O(\Delta t^2).$$

Using the relation (3.37) in the above equation and then dividing by Δt gives

$$s\Delta t(u_l - u_r) = \Delta t(f(u_l) - f(u_r)) + O(\Delta t)$$

where s, u_l, and u_r are all evaluated at t_1. Letting $\Delta t \to 0$ gives the R-H condition (3.33).

EXERCISE 3.8. *Solve Burgers' equation with initial data*

$$u_0(x) = \begin{cases} 2 & x < 0 \\ 1 & 0 < x < 2 \\ 0 & x > 2. \end{cases} \tag{3.39}$$

Sketch the characteristics and shock paths in the x-t plane. Hint: The two shocks merge into one shock at some point.

The equal area rule. One technique that is useful for determining weak solutions by hand is to start with the "solution" constructed using characteristics (which may be multi-valued if characteristics cross) and then eliminate the multi-valued parts by inserting shocks. The shock location can be determined by the "equal area rule", which is best understood by looking at Figure 3.13. The shock is located such that the shaded regions cut off on either side have equal areas, as in Figure 3.13b. This is a consequence of conservation — the integral of the discontinuous weak solution (shaded area in Figure 3.13c) must be the same as the area "under" the multi-valued solution (shaded area in 3.13a), since both "solve" the same conservation law.

3.7 Manipulating conservation laws

One danger to observe in dealing with conservation laws is that transforming the differential form into what appears to be an equivalent differential equation may not give an equivalent equation in the context of weak solutions.

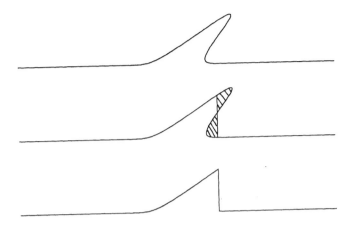

Figure 3.13. Equal area rule for shock location.

EXAMPLE 3.3. If we multiply Burgers' equation

$$u_t + \left(\frac{1}{2}u^2\right)_x = 0 \tag{3.40}$$

by $2u$, we obtain $2uu_t + 2u^2u_x = 0$, which can be rewritten as

$$(u^2)_t + \left(\frac{2}{3}u^3\right)_x = 0. \tag{3.41}$$

This is again a conservation law, now for u^2 rather than u itself, with flux function $f(u^2) = \frac{2}{3}(u^2)^{3/2}$. The differential equations (3.40) and (3.41) have precisely the same smooth solutions. However, they have different weak solutions, as we can see by considering the Riemann problem with $u_l > u_r$. The unique weak solution of (3.40) is a shock traveling at speed

$$s_1 = \frac{\left[\frac{1}{2}u^2\right]}{[u]} = \frac{1}{2}(u_l + u_r), \tag{3.42}$$

whereas the unique weak solution to (3.41) is a shock traveling at speed

$$s_2 = \frac{\left[\frac{2}{3}u^3\right]}{[u^2]} = \frac{2}{3}\left(\frac{u_r^3 - u_l^3}{u_r^2 - u_l^2}\right). \tag{3.43}$$

It is easy to check that

$$s_2 - s_1 = \frac{1}{6}\frac{(u_l - u_r)^2}{(u_l + u_r)} \tag{3.44}$$

and so $s_2 \neq s_1$ when $u_l \neq u_r$, and the two equations have different weak solutions. The derivation of (3.41) from (3.40) requires manipulating derivatives in a manner that is valid only when u is smooth.

3.8 Entropy conditions

As demonstrated above, there are situations in which the weak solution is not unique and an additional condition is required to pick out the physically relevant vanishing viscosity solution. The condition which defines this solution is that it should be the limiting solution of the viscous equation as $\epsilon \to 0$, but this is not easy to work with. We want to find simpler conditions.

For scalar equations there is an obvious condition suggested by Figures 3.8 and 3.10. A shock should have characteristics going *into* the shock, as time advances. A propagating discontinuity with characteristics coming *out* of it, as in Figure 3.9, is unstable to perturbations. Either smearing out the initial profile a little, or adding some viscosity to the system, will cause this to be replaced by a rarefaction fan of characteristics, as in Figure 3.10. This gives our first version of the entropy condition:

ENTROPY CONDITION (VERSION I): *A discontinuity propagating with speed s given by (3.33) satisfies the entropy condition if*

$$f'(u_l) > s > f'(u_r). \tag{3.45}$$

Note that $f'(u)$ is the characteristic speed. For convex f, the Rankine-Hugoniot speed s from (3.34) must lie between $f'(u_l)$ and $f'(u_r)$, so (3.45) reduces to simply the requirement that $f'(u_l) > f'(u_r)$, which again by convexity requires $u_l > u_r$.

A more general form of this condition, due to Oleinik, applies also to nonconvex scalar flux functions f:

ENTROPY CONDITION (VERSION II): *$u(x,t)$ is the entropy solution if all discontinuities have the property that*

$$\frac{f(u) - f(u_l)}{u - u_l} \geq s \geq \frac{f(u) - f(u_r)}{u - u_r} \tag{3.46}$$

for all u between u_l and u_r.

For convex f, this requirement reduces to (3.45).

Another form of the entropy condition is based on the spreading of characteristics in a rarefaction fan. If $u(x,t)$ is an increasing function of x in some region, then characteristics spread out if $f'' > 0$. The rate of spreading can be quantified, and gives the following condition, also due to Oleinik[57].

ENTROPY CONDITION (VERSION III): *$u(x,t)$ is the entropy solution if there is a constant $E > 0$ such that for all $a > 0$, $t > 0$ and $x \in \mathbb{R}$,*

$$\frac{u(x + a, t) - u(x, t)}{a} < \frac{E}{t}. \tag{3.47}$$

Note that for a discontinuity propagating with constant left and right states u_l and u_r, this can be satisfied only if $u_r - u_l \leq 0$, so this agrees with (3.45). The form of (3.47) may seem unnecessarily complicated, but it turns out to be easier to apply in some contexts. In particular, this formulation has advantages in studying numerical methods. One problem we face in developing numerical methods is guaranteeing that they converge to the correct solution. Some numerical methods are known to converge to the *wrong* weak solution in some instances. The criterion (3.45) is hard to apply to discrete solutions — a discrete approximation defined only at grid points is in some sense discontinuous everywhere. If $U_j < U_{j+1}$ at some grid point, how do we determine whether this is a jump that violates the entropy condition, or is merely part of a smooth approximation of a rarefaction wave? Intuitively, we know the answer: it's part of a smooth approximation, and therefore acceptable, if the size of this jump is $O(\Delta x)$ as we refine the grid and $\Delta x \to 0$. To turn this into a proof that some method converges to the correct solution, we must quantify this requirement and (3.47) gives what we need. Taking $a = \Delta x$, we must ensure that there is a constant $E > 0$ such that

$$U_{j+1}(t) - U_j(t) < \left(\frac{E}{t}\right) \Delta x \tag{3.48}$$

for all $t > 0$ as $\Delta x \to 0$. This inequality can often be established for discrete methods.

In fact, Oleinik's original proof that an entropy solution to (3.13) satisfying (3.47) always exists proceeds by defining such a discrete approximation and then taking limits. This is also presented in Theorem 16.1 of Smoller[77].

3.8.1 Entropy functions

Yet another approach to the entropy condition is to define an entropy function $\eta(u)$ for which an additional conservation law holds for smooth solutions that becomes an inequality for discontinuous solutions. In gas dynamics, there exists a physical quantity called the entropy that is known to be constant along particle paths in smooth flow and to jump to a higher value as the gas crosses a shock. It can never jump to a lower value, and this gives the physical entropy condition that picks out the correct weak solution in gas dynamics.

Suppose some function $\eta(u)$ satisfies a conservation law of the form

$$\eta(u)_t + \psi(u)_x = 0 \tag{3.49}$$

for some **entropy flux** $\psi(u)$. Then we can obtain from this, for smooth u,

$$\eta'(u)u_t + \psi'(u)u_x = 0. \tag{3.50}$$

Recall that the conservation law (3.13) can be written as $u_t + f'(u)u_x = 0$. Multiply this by $\eta'(u)$ and compare with (3.50) to obtain

$$\psi'(u) = \eta'(u)f'(u). \tag{3.51}$$

For a scalar conservation law this equation admits many solutions $\eta(u)$, $\psi(u)$. For a system of equations η and ψ are still *scalar* functions, but now (3.51) reads $\nabla\psi(u) = f'(u)\nabla\eta(u)$, which is a system of m equations for the two variables η and ψ. If $m > 2$ this may have no solutions.

An additional condition we place on the entropy function is that it be *convex*, $\eta''(u) > 0$, for reasons that will be seen below.

The entropy $\eta(u)$ is conserved for *smooth* flows by its definition. For discontinuous solutions, however, the manipulations performed above are not valid. Since we are particularly interested in how the entropy behaves for the vanishing viscosity weak solution, we look at the related viscous problem and will then let the viscosity tend to zero. The viscous equation is

$$u_t + f(u)_x = \epsilon u_{xx}. \tag{3.52}$$

Since solutions to this equation are always smooth, we can derive the corresponding evolution equation for the entropy following the same manipulations we used for smooth solutions of the inviscid equation, multiplying (3.52) by $\eta'(u)$ to obtain

$$\eta(u)_t + \psi(u)_x = \epsilon \eta'(u)u_{xx}. \tag{3.53}$$

We can now rewrite the right hand side to obtain

$$\eta(u)_t + \psi(u)_x = \epsilon(\eta'(u)u_x)_x - \epsilon\eta''(u)u_x^2. \tag{3.54}$$

Integrating this equation over the rectangle $[x_1, x_2] \times [t_1, t_2]$ gives

$$\int_{t_1}^{t_2}\int_{x_1}^{x_2} \eta(u)_t + \psi(u)_x \, dx \, dt = \epsilon \int_{t_1}^{t_2} [\eta'(u(x_2, t))u_x(x_2, t) - \eta'(u(x_1, t))u_x(x_1, t)] \, dt$$
$$- \epsilon \int_{t_1}^{t_2}\int_{x_1}^{x_2} \eta''(u)u_x^2 \, dx \, dt.$$

As $\epsilon \to 0$, the first term on the right hand side vanishes. (This is clearly true if u is smooth at x_1 and x_2, and can be shown more generally.) The other term, however, involves integrating u_x^2 over the $[x_1, x_2] \times [t_1, t_2]$. If the limiting weak solution is discontinuous along a curve in this rectangle, then this term will not vanish in the limit. However, since $\epsilon > 0$, $u_x^2 > 0$ and $\eta'' > 0$ (by our convexity assumption), we can conclude that the right hand side is nonpositive in the limit and hence the vanishing viscosity weak solution satisfies

$$\int_{t_1}^{t_2}\int_{x_1}^{x_2} \eta(u)_t + \psi(u)_x \, dx \, dt \leq 0 \tag{3.55}$$

for all x_1, x_2, t_1 and t_2. Alternatively, in integral form,

$$\int_{x_1}^{x_2} \eta(u(x,t))\, dx \bigg|_{t_1}^{t_2} + \int_{t_1}^{t_2} \psi(u(x,t))\, dt \bigg|_{x_1}^{x_2} \leq 0, \tag{3.56}$$

i.e.,

$$\int_{x_1}^{x_2} \eta(u(x,t_2))\, dx \ \leq \ \int_{x_1}^{x_2} \eta(u(x,t_1))\, dx \tag{3.57}$$
$$- \left(\int_{t_1}^{t_2} \psi(u(x_2,t))\, dt - \int_{t_1}^{t_2} \psi(u(x_1,t))\, dt \right).$$

Consequently, the total integral of η is not necessarily conserved, but can only *decrease*. (Note that our mathematical assumption of convexity leads to an "entropy function" that decreases, whereas the physical entropy in gas dynamics increases.) The fact that (3.55) holds for all x_1, x_2, t_1 and t_2 is summarized by saying that $\eta(u)_t + \psi(u)_x \leq 0$ in the weak sense. This gives our final form of the entropy condition, called the **entropy inequality**.

ENTROPY CONDITION (VERSION IV): *The function $u(x,t)$ is the entropy solution of (3.13) if, for all convex entropy functions and corresponding entropy fluxes, the inequality*

$$\eta(u)_t + \psi(u)_x \leq 0 \tag{3.58}$$

is satisfied in the weak sense.

This formulation is also useful in analyzing numerical methods. If a discrete form of this entropy inequality is known to hold for some numerical method, then it can be shown that the method converges to the entropy solution.

Just as for the conservation law, an alternative weak form of the entropy condition can be formulated by integrating against smooth test functions ϕ, now required to be nonnegative since the entropy condition involves an inequality. The **weak form of the entropy inequality** is

$$\int_0^\infty \int_{-\infty}^\infty \phi_t(x,t)\eta(u(x,t)) + \phi_x(x,t)\psi(u(x,t))\, dx\, dt$$
$$\leq - \int_{-\infty}^\infty \phi(x,0)\eta(u(x,0))\, dx \tag{3.59}$$

for all $\phi \in C_0^1(\mathbb{R} \times \mathbb{R})$ with $\phi(x,t) \geq 0$ for all x, t.

EXAMPLE 3.4.   Consider Burgers' equation with $f(u) = \frac{1}{2}u^2$ and take $\eta(u) = u^2$. Then (3.51) gives $\psi'(u) = 2u^2$ and hence $\psi(u) = \frac{2}{3}u^3$. Then entropy condition (3.58) reads

$$(u^2)_t + \left(\frac{2}{3}u^3 \right)_x \leq 0. \tag{3.60}$$

For smooth solutions of Burgers' equation this should hold with equality, as we have already seen in Example 3.3. If a discontinuity is present, then integrating $(u^2)_t + (\frac{2}{3}u^3)_x$ over an infinitesmal rectangle as in Figure 3.12 gives

$$\int_{x_1}^{x_2} u^2(x,t)\,dx \bigg|_{t_1}^{t_2} + \int_{t_1}^{t_2} \frac{2}{3}u^3(x,t)\,dt \bigg|_{x_1}^{x_2} = s_1 \Delta t(u_l^2 - u_r^2) + \frac{2}{3}\Delta t(u_r^3 - u_l^3) + O(\Delta t^2)$$
$$= \Delta t(u_l^2 - u_r^2)(s_1 - s_2) + O(\Delta t^2)$$
$$= -\frac{1}{6}(u_l - u_r)^3 \Delta t + O(\Delta t^2)$$

where s_1 and s_2 are given by (3.42) and (3.43) and we have used (3.44). For small $\Delta t > 0$, the $O(\Delta t^2)$ term will not affect the sign of this quantity and so the weak form (3.56) is satisfied if and only if $(u_l - u_r)^3 > 0$, and hence the only allowable discontinuities have $u_l > u_r$, as expected.

4 Some Scalar Examples

In this chapter we will look at a couple of examples of scalar conservation laws with some physical meaning, and apply the theory developed in the previous chapter. The first of these examples (traffic flow) should also help develop some physical intuition that is applicable to the more complicated case of gas dynamics, with gas molecules taking the place of cars. This application is discussed in much more detail in Chapter 3 of Whitham[97]. The second example (two phase flow) shows what can happen when f is not convex.

4.1 Traffic flow

Consider the flow of cars on a highway. Let ρ denote the density of cars (in vehicles per mile, say) and u the velocity. In this application ρ is restricted to a certain range, $0 \le \rho \le \rho_{max}$, where ρ_{max} is the value at which cars are bumper to bumper.

Since cars are conserved, the density and velocity must be related by the continuity equation derived in Section 1,

$$\rho_t + (\rho u)_x = 0. \tag{4.1}$$

In order to obtain a scalar conservation law for ρ alone, we now assume that u is a given function of ρ. This makes sense: on a highway we would optimally like to drive at some speed u_{max} (the speed limit perhaps), but in heavy traffic we slow down, with velocity decreasing as density increases. The simplest model is the linear relation

$$u(\rho) = u_{max}(1 - \rho/\rho_{max}). \tag{4.2}$$

At zero density (empty road) the speed is u_{max}, but decreases to zero as ρ approaches ρ_{max}. Using this in (4.1) gives

$$\rho_t + f(\rho)_x = 0 \tag{4.3}$$

where

$$f(\rho) = \rho u_{max}(1 - \rho/\rho_{max}). \tag{4.4}$$

41

Whitham notes that a good fit to data for the Lincoln tunnel was found by Greenberg in 1959 by

$$f(\rho) = a\rho \log(\rho_{max}/\rho),$$

a function shaped similar to (4.4).

The characteristic speeds for (4.3) with flux (4.4) are

$$f'(\rho) = u_{max}(1 - 2\rho/\rho_{max}), \tag{4.5}$$

while the shock speed for a jump from ρ_l to ρ_r is

$$s = \frac{f(\rho_l) - f(\rho_r)}{\rho_l - \rho_r} = u_{max}(1 - (\rho_l + \rho_r)/\rho_{max}). \tag{4.6}$$

The entropy condition (3.45) says that a propagating shock must satisfy $f'(\rho_l) > f'(\rho_r)$ which requires $\rho_l < \rho_r$. Note this is the opposite inequality as in Burgers' equation since here f is concave rather than convex.

We now consider a few examples of solutions to this equation and their physical interpretation.

EXAMPLE 4.1. Take initial data

$$\rho(x, 0) = \begin{cases} \rho_l & x < 0 \\ \rho_r & x > 0 \end{cases} \tag{4.7}$$

where $0 < \rho_l < \rho_r < \rho_{max}$. Then the solution is a shock wave traveling with speed s given by (4.6). Note that although $u(\rho) \geq 0$ the shock speed s can be either positive or negative depending on ρ_l and ρ_r.

Consider the case $\rho_r = \rho_{max}$ and $\rho_l < \rho_{max}$. Then $s < 0$ and the shock propagates to the left. This models the situation in which cars moving at speed $u_l > 0$ unexpectedly encounter a bumper-to-bumper traffic jam and slam on their brakes, instantaneously reducing their velocity to 0 while the density jumps from ρ_l to ρ_{max}. This discontinuity occurs at the shock, and clearly the shock location moves to the left as more cars join the traffic jam. This is illustrated in Figure 4.1, where the vehicle trajectories ("particle paths") are sketched. Note that the distance between vehicles is inversely proportional to density. (In gas dynamics, $1/\rho$ is called the *specific volume*.)

The particle paths should not be confused with the characteristics, which are shown in Figure 4.2 for the case $\rho_l = \frac{1}{2}\rho_{max}$ (so $u_l = \frac{1}{2}u_{max}$), as is the case in Figure 4.1 also. In this case, $f'(\rho_l) = 0$. If $\rho_l > \frac{1}{2}\rho_{max}$ then $f'(\rho_l) < 0$ and all characteristics go to the left, while if $\rho_l < \frac{1}{2}\rho_{max}$ then $f'(\rho_l) > 0$ and characteristics to the left of the shock are rightward going.

EXERCISE 4.1. *Sketch the particle paths and characteristics for a case with $\rho_l + \rho_r < \rho_{max}$.*

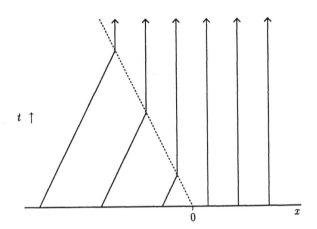

Figure 4.1. Traffic jam shock wave (vehicle trajectories), with data $\rho_l = \frac{1}{2}\rho_{max}$, $\rho_r = \rho_{max}$.

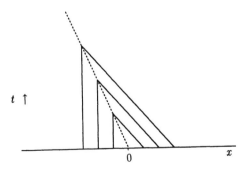

Figure 4.2. Characteristics.

EXAMPLE 4.2. Again consider a Riemann problem with data of the form (4.7), but now take $0 < \rho_r < \rho_l < \rho_{max}$ so that the solution is a rarefaction wave. Figure 4.3 shows the case where $\rho_l = \rho_{max}$ and $\rho_r = \frac{1}{2}\rho_{max}$. This might model the startup of cars after a light turns green. The cars to the left are initially stationary but can begin to accelerate once the cars in front of them begin to move. Since the velocity is related to the density by (4.2), each driver can speed up only by allowing the distance between her and the previous car to increase, and so we see a gradual acceleration and spreading out of cars.

As cars go through the rarefaction wave, the density decreases. Cars spread out or become "rarefied" in the terminology used for gas molecules.

Of course in this case there is another weak solution to (4.3), the entropy-violating shock. This would correspond to drivers accelerating instantaneously from $u_l = 0$ to $u_r > 0$ as the preceding car moves out of the way. This behavior is not usually seen in practice except perhaps in high school parking lots.

The reason is that in practice there is "viscosity", which here takes the form of slow response of drivers and automobiles. In the shock wave example above, the instantaneous jump from $u_l > 0$ to $u_r = 0$ as drivers slam on their brakes is obviously a mathematical idealization. However, in terms of modeling the big picture — how the traffic jam evolves — the detailed structure of $u(x)$ in the shock is unimportant.

EXERCISE 4.2. *For cars starting at a green light with open road ahead of them, the initial conditions would really be (4.7) with $\rho_l = \rho_{max}$ and $\rho_r = 0$. Solve this Riemann problem and sketch the particle paths and characteristics.*

EXERCISE 4.3. *Sketch the distribution of ρ and u at some fixed time $t > 0$ for the solution of Exercise 4.2.*

EXERCISE 4.4. *Determine the manner in which a given car accelerates in the solution to Exercise 4.2, i.e. determine $v(t)$ where v represents the velocity along some particular particle path as time evolves.*

4.1.1   Characteristics and "sound speed"

For a scalar conservation law, information always travels with speed $f'(\rho)$ as long as the solution is smooth. If fact, the solution is constant along characteristics since

$$\rho_t + f'(\rho)\rho_x = 0. \tag{4.8}$$

We can obtain another interpretation of this if we consider the special case of nearly constant initial data, say

$$\rho(x,0) = \rho_0 + \epsilon\rho_1(x,0). \tag{4.9}$$

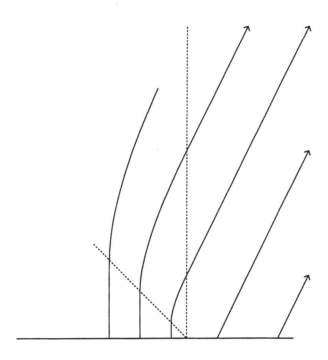

Figure 4.3. Rarefaction wave (vehicle trajectories), with data $\rho_l = \rho_{max}$, $\rho_r = \frac{1}{2}\rho_{max}$.

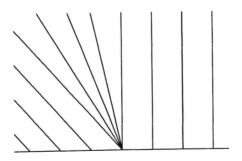

Figure 4.4. Characteristics.

Then we can approximate our nonlinear equation by a linear equation. Assuming

$$\rho(x,t) = \rho_0 + \epsilon\rho_1(x,t) \tag{4.10}$$

remains valid with $\rho_1 = O(1)$, we find that

$$\rho_t = \epsilon\rho_{1t}$$
$$\rho_x = \epsilon\rho_{1x}$$
$$f'(\rho) = f'(\rho_0) + \epsilon\rho_1 f''(\rho_0) + O(\epsilon^2).$$

Using these in (4.8) and dividing by ϵ gives

$$\rho_{1t} + f'(\rho_0)\rho_{1x} = -\epsilon f''(\rho_0)\rho_1\rho_{1x} + O(\epsilon^2). \tag{4.11}$$

For small ϵ the behavior, at least for times $t \ll 1/\epsilon$, is governed by the equation obtained by ignoring the right hand side. This gives a constant coefficient linear advection equation for $\rho_1(x,t)$:

$$\rho_{1t} + f'(\rho_0)\rho_{1x} = 0. \tag{4.12}$$

The initial data simply propagates unchanged with velocity $f'(\rho_0)$.

In the traffic flow model this corresponds to a situation where cars are nearly evenly spaced with small variation in the density. These variations will propagate with velocity roughly $f'(\rho_0)$.

As a specific example, suppose the data is constant except for a small rise in density at some point, a crowding on the highway. The cars in this disturbance are going slower than the cars either ahead or behind, with two effects. First, since they are going slower than the cars behind, the cars behind will start to catch up, seeing a rise in their local density and therefore be forced to slow down. Second, since the cars in the disturbance are going slower than the cars ahead of them, they will start to fall behind, leading to a decrease in their local density and an increase in speed. The consequence is that the disturbance will propagate "backwards" through the line of cars. Here by "backwards" I mean from the standpoint of any given driver. He slows down because the cars in front of him have, and his behavior in turn affects the drivers behind him.

Note that in spite of this, the speed at which the disturbance propagates could be *positive*, if $f'(\rho_0) > 0$, which happens if $\rho_0 < \frac{1}{2}\rho_{max}$. This is illustrated in Figure 4.5. Here the vehicle trajectories are sketched. The jog in each trajectory is the effect of the car slowing down as the disturbance passes.

The nearly linear behavior of small amplitude disturbances is also seen in gas dynamics. In fact, this is precisely how sound waves propagate. If the gas is at rest, $v_0 = 0$ in the linearization, then sound propagates at velocities $\pm c$, where the **sound speed** c depends on the equation of state (we will see that c^2 is the derivative of pressure with respect to density at constant entropy). If we add some uniform motion to the gas, so $v_0 \neq 0$, then

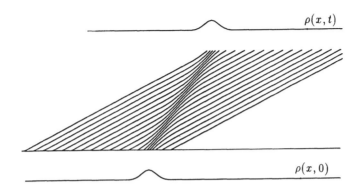

$\rho(x, t)$

$\rho(x, 0)$

Figure 4.5. Vehicle trajectories and propagation of a small disturbance.

sound waves propagate at speeds $v_0 \pm c$. This simple shift arises from the fact that you can add a uniform velocity to a solution of the Euler equations and it remains a solution.

This is not true for the traffic flow model, since u is assumed to be a given function of ρ. However, it should be be clear that the velocity which corresponds most naturally to the sound speed in gas dynamics is

$$c = f'(\rho_0) - u(\rho_0), \tag{4.13}$$

so that disturbances propagate at speed $f'(\rho_0) = u(\rho_0) + c$, or at speed c relative to the traffic flow. Using (4.2) and (4.5), this becomes

$$c = -u_{max}\, \rho_0/\rho_{max}. \tag{4.14}$$

EXERCISE 4.5. *What is the physical significance of the fact that $c < 0$?*

In gas dynamics the case $v < c$ is called **subsonic flow**, while if $v > c$ the flow is **supersonic**. By analogy, the value $\rho_0 = \frac{1}{2}\rho_{max}$ at which $f'(\rho_0) = 0$ is called the **sonic point**, since this is the value for which $u(\rho_0) = c$. For $\rho < \frac{1}{2}\rho_{max}$, the cars are moving faster than disturbances propagate backwards through the traffic, giving the situation already illustrated in Figure 4.5.

EXERCISE 4.6. *Sketch particle paths similar to Figure 4.5 for the case $\rho_0 = \frac{1}{2}\rho_{max}$.*

EXERCISE 4.7. *Consider a shock wave with left and right states ρ_l and ρ_r, and let the shock strength approach zero, by letting $\rho_l \to \rho_r$. Show that the shock speed for these weak shocks approaches the linearized propagation speed $f'(\rho_r)$.*

f(u)

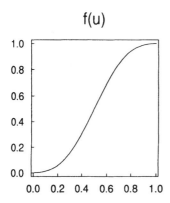

f'(u)

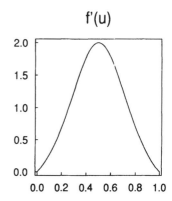

Figure 4.6. Flux function for Buckley-Leverett equation.

4.2 Two phase flow

When f is convex, the solution to the Riemann problem is always either a shock or a rarefaction wave. When f is not convex, the entropy solution might involve both. To illustrate this, we will look at the **Buckley-Leverett equations**, a simple model for two phase fluid flow in a porous medium. One application is to oil reservoir simulation. When an underground source of oil is tapped, a certain amount of oil flows out on its own due to high pressure in the reservoir. After the flow stops, there is typically a large amount of oil still in the ground. One standard method of "secondary recovery" is to pump water into the oil field through some wells, forcing oil out through others. In this case the two phases are oil and water, and the flow takes place in a porous medium of rock or sand.

The Buckley-Leverett equations are a particularly simple scalar model that captures some features of this flow. In one space dimension the equation has the standard conservation law form (3.13) with

$$f(u) = \frac{u^2}{u^2 + a(1-u)^2} \tag{4.15}$$

where a is a constant. Figure 4.6 shows $f(u)$ when $a = 1/2$. Here u represents the saturation of water and so lies between 0 and 1.

Now consider the Riemann problem with initial states $u_l = 1$ and $u_r = 0$, modeling the flow of pure water ($u = 1$) into pure oil ($u = 0$). By following characteristics, we can construct the triple-valued solution shown in Figure 4.7a. Note that the characteristic velocities are $f'(u)$ so that the profile of this bulge seen here at time t is simply the graph of $tf'(u)$ turned sideways.

We can now use the equal area rule to replace this triple-valued solution by a correct shock. The resulting weak solution is shown in Figure 4.7b, along with the characteristics in Figure 4.7c.

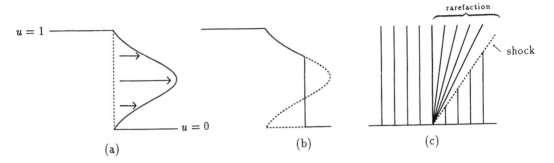

Figure 4.7. Riemann solution for Buckley-Leverett equation.

EXERCISE 4.8. *Use the equal area rule to find an expression for the shock location as a function of t and verify that the Rankine-Hugoniot condition is always satisfied.*

If you do the above exercise, you will find that the shock location moves at a constant speed, and the post-shock value u^* is also constant. This might be surprising, unless you are familiar with self-similarity of solutions to the Riemann problem, in which case you should have expected this. This will be discussed later.

Note the physical interpretation of the solution shown in Figure 4.7. As the water moves in, it displaces a certain fraction u^* of the oil immediately. Behind the shock, there is a mixture of oil and water, with less and less oil as time goes on. At a production well (at the point $x = 1$, say), one obtains pure oil until the shock arrives, followed by a mixture of oil and water with diminishing returns as time goes on. It is impossible to recover all of the oil in finite time by this technique.

Note that the Riemann solution involves both a shock and a rarefaction wave. If $f(u)$ had more inflection points, the solution might involve several shocks and rarefactions.

EXERCISE 4.9. *Explain why it is impossible to have a Riemann solution involving both a shock and a rarefaction when f is convex or concave.*

It turns out that the solution to the Riemann problem can be determined from the graph of f in a simple manner. If $u_r < u_l$ then construct the **convex hull** of the set $\{(x, y) : u_r \leq x \leq u_l$ and $y \leq f(x)\}$. The convex hull is the smallest convex set containing the original set. This is shown in Figure 4.8 for the case $u_l = 1$, $u_r = 0$.

If we look at the upper boundary of this set, we see that it is composed of a straight line segment from $(0,0)$ to $(u^*, f(u^*))$ and then follows $y = f(x)$ up to $(1,1)$. The point of tangency u^* is precisely the post-shock value. The straight line represents a shock jumping from $u = 0$ to $u = u^*$ and the segment where the boundary follows $f(x)$ is the

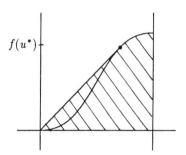

Figure 4.8. Convex hull showing shock and rarefaction.

rarefaction wave. This works more generally for any two states (provided $u_l > u_r$) and for any f.

Note that the slope of the line segment is $s^* = [f(u^*) - f(u_r)] / [u^* - u_r]$, which is precisely the shock speed. The fact that this line is tangent to the curve $f(x)$ at u^* means that $s^* = f'(u^*)$, the shock moves at the same speed as the characteristics at this edge of the rarefaction fan, as seen in Figure 4.7c.

If the shock were connected to some point $\hat{u} < u^*$, then the shock speed $[f(\hat{u}) - f(u_r)] / [\hat{u} - u_r]$ would be less than $f'(\hat{u})$, leading to a triple-valued solution. On the other hand, if the shock were connected to some point above u^* then the entropy condition (3.46) would be violated. This explains the tangency requirement, which comes out naturally from the convex hull construction.

EXERCISE 4.10. *Show that (3.46) is violated if the shock goes above u^*.*

If $u_l < u_r$ then the same idea works but we look instead at the convex hull of the set of points *above* the graph, $\{(x, y): u_l \le x \le u_r$ and $y \ge f(x)\}$.

Note that if f is convex, then the convex hull construction gives either a single line segment (single shock) if $u_l > u_r$ or the function f itself (single rarefaction) if $u_l < u_r$.

5 Some Nonlinear Systems

Before developing the theory for systems of conservation laws, it is useful to have some specific examples in mind. In this chapter we will derive some systems of conservation laws.

5.1 The Euler equations

The Euler equations of gas dynamics are a particularly important example. The continuity equation for conservation of mass was derived in Chapter 2. Here we will consider the momentum and energy equations in more detail, as well as the equation of state and a few other quantities of physical (and mathematical) significance, such as the entropy. We will also look at some simplifications, the isentropic and isothermal cases, where systems of two equations are obtained. These provide very good examples which will be used in the coming chapters to illustrate the nonlinear theory.

The derivations here will be very brief, with an emphasis on the main ideas without a detailed description of the physics. A more thorough introduction can be found in Whitham[97], Courant-Friedrichs[11], or any good book on gas dynamics, e.g. [51],[71],[94].

Recall that ρ is the density, v the velocity, E the total energy, and p the pressure of the gas. The continuity equation is

$$\rho_t + (\rho v)_x = 0. \tag{5.1}$$

The mass flux is given by ρv. More generally, for any quantity z that is advected with the flow there will be a contribution to the flux for z of the form zv. Thus, the momentum equation has a contribution of the form $(\rho v)v = \rho v^2$ and the energy equation has a flux contribution Ev.

In addition to advection, there are forces on the fluid that cause acceleration due to Newton's laws, and hence changes in momentum. If there are no outside forces, then the only force is due to variations in the fluid itself, and is proportional to the pressure gradient which is simply p_x in one dimension. Combining this with the advective flux gives the momentum equation

$$(\rho v)_t + (\rho v^2 + p)_x = 0. \tag{5.2}$$

The total energy E is often decomposed as

$$E = \frac{1}{2}\rho v^2 + \rho e. \tag{5.3}$$

The first term here is the kinetic energy, while ρe is the internal energy. The variable e, internal energy per unit mass, is called the **specific internal energy**. (In general "specific" means "per unit mass"). Internal energy includes rotational and vibrational energy and possibly other forms of energy in more complicated situations. In the Euler equations we assume that the gas is in chemical and thermodynamic equilibrium and that the internal energy is a known function of pressure and density:

$$e = e(p, \rho). \tag{5.4}$$

This is the "equation of state" for the gas, which depends on the particular gas under study.

The total energy advects with the flow, but is also modified due to work done on the system. In the absence of outside forces, work is done only by the pressure forces and is proportional to the gradient of vp. The conservation law for total energy thus takes the form

$$E_t + [v(E + p)]_x = 0 \tag{5.5}$$

in one dimension.

Putting these equations together gives the system of Euler equations

$$\begin{bmatrix} \rho \\ \rho v \\ E \end{bmatrix}_t + \begin{bmatrix} \rho v \\ \rho v^2 + p \\ v(E + p) \end{bmatrix}_x = 0. \tag{5.6}$$

In two space dimensions the Euler equations take the form

$$\begin{bmatrix} \rho \\ \rho u \\ \rho v \\ E \end{bmatrix}_t + \begin{bmatrix} \rho u \\ \rho u^2 + p \\ \rho uv \\ u(E + p) \end{bmatrix}_x + \begin{bmatrix} \rho v \\ \rho uv \\ \rho v^2 + p \\ v(E + p) \end{bmatrix}_y = 0. \tag{5.7}$$

where (u, v) is the 2D fluid velocity.

The equation of state. We still need to specify the equation of state relating the internal energy to pressure and density.

5.1.1 Ideal gas

For an ideal gas, internal energy is a function of temperature alone, $e = e(T)$, and T is related to p and ρ by the **ideal gas law**,

$$p = \mathcal{R}\rho T \tag{5.8}$$

where $\mathcal{R}$ is a constant. To good approximation, the internal energy is simply proportional to the temperature,

$$e = c_v T, \tag{5.9}$$

where c_v is a constant known as the **specific heat at constant volume**. Such gases are called **polytropic**. If energy is added to a fixed quantity of gas, and the volume is held constant, then the change in energy and change in temperature are related via

$$de = c_v dT. \tag{5.10}$$

On the other hand, if the gas is allowed to expand while the energy is added, and pressure is held constant instead, not all of the energy goes into increasing the internal energy. The work done in expanding the volume $1/\rho$ by $d(1/\rho)$ is $pd(1/\rho)$ and we obtain another relation

$$de + pd(1/\rho) = c_p dT \tag{5.11}$$

or

$$d(e + p/\rho) = c_p dT \tag{5.12}$$

where c_p is the **specific heat at constant pressure**. The quantity

$$h = e + p/\rho \tag{5.13}$$

is called the **enthalpy**. For a polytropic gas, c_p is also assumed to be constant so that (5.12) yields

$$h = c_p T. \tag{5.14}$$

Note that by the ideal gas law,

$$c_p - c_v = \mathcal{R}. \tag{5.15}$$

The equation of state for an polytropic gas turns out to depend only on the **ratio of specific heats**, usually denoted by

$$\gamma = c_p / c_v. \tag{5.16}$$

Internal energy in a molecule is typically split up between various degrees of freedom (translational, rotational, vibrational, etc.). How many degrees of freedom exist depends on the nature of the gas. The general *principle of equipartition of energy* says that the average energy in each of these is the same. Each degree of freedom contributes an average energy of $\frac{1}{2}kT$ per molecule, where k is **Boltzmann's constant**. This gives a total contribution of $\frac{\alpha}{2}kT$ per molecule if there are α degrees of freedom. Multiplying this by n, the number of molecules per unit mass (which depends on the gas), gives

$$e = \frac{\alpha}{2}nkT. \tag{5.17}$$

The product nk is precisely the gas constant $\mathcal{R}$, so comparing this to (5.9) gives

$$c_v = \frac{\alpha}{2}\mathcal{R}. \tag{5.18}$$

From (5.15) we obtain

$$c_p = \left(1 + \frac{\alpha}{2}\right)\mathcal{R}, \tag{5.19}$$

and so

$$\gamma = c_p/c_v = \frac{\alpha + 2}{\alpha}. \tag{5.20}$$

For a monatomic gas the only degrees of freedom are the three translational degrees, so $\alpha = 3$ and $\gamma = 5/3$. For a diatomic gas (such as air, which is composed primarily of N_2 and O_2), there are also two rotational degrees of freedom and $\alpha = 5$, so that $\gamma = 7/5 = 1.4$.

The equation of state for a polytropic gas. Note that $T = p/\mathcal{R}\rho$ so that

$$e = c_v T = \left(\frac{c_v}{\mathcal{R}}\right)\frac{p}{\rho} = \frac{p}{(\gamma - 1)\rho} \tag{5.21}$$

by (5.15) and (5.16). Using this in (5.3) gives the common form of the equation of state for a polytropic gas:

$$E = \frac{p}{\gamma - 1} + \frac{1}{2}\rho v^2. \tag{5.22}$$

5.1.2 Entropy

Another important thermodynamic quantity is the entropy. Roughly speaking, this measures the disorder in the system. The entropy S is defined up to an additive constant by

$$S = c_v \log(p/\rho^\gamma) + \text{constant}. \tag{5.23}$$

This can be solved for p to give

$$p = \kappa e^{S/c_v}\rho^\gamma, \tag{5.24}$$

where κ is a constant.

From the Euler equations we can derive the relation

$$S_t + vS_x = 0 \tag{5.25}$$

which says that entropy is constant along particle paths in regions of smooth flow. In fact, (5.25) can be derived from fundamental principles and this equation, together with the conservation of mass and momentum equations, gives an alternative formulation of the Euler equations (though not in conservation form):

$$\begin{aligned} \rho_t + (\rho v)_x &= 0 \\ (\rho v)_t + (\rho v^2 + p)_x &= 0 \\ S_t + v S_x &= 0 \end{aligned} \qquad (5.26)$$

It turns out that the equation of state then gives p as a function of ρ and S alone, e.g. (5.24) for a polytropic gas. In this form the partial derivative of p with respect to ρ (holding S fixed) plays a fundamental role: its square root c is the local **speed of sound** in the gas. For a polytropic gas we have

$$c^2 = \left.\frac{\partial p}{\partial \rho}\right|_{S=\text{constant}} = \gamma \kappa e^{S/c_v} \rho^{\gamma - 1} = \gamma p / \rho \qquad (5.27)$$

and so

$$c = \sqrt{\gamma p / \rho}. \qquad (5.28)$$

From our standpoint the most important property of entropy is the fact that in smooth flow entropy remains constant on each particle path, while if a particle crosses a shock then the entropy may jump, but must *increase*. This is the physical entropy condition for shocks.

Note that along a particle path in smooth flow, since S is constant we find by (5.24) that

$$p = \hat{\kappa} \rho^\gamma \qquad (5.29)$$

where $\hat{\kappa} = \kappa e^{S/c_v}$ is a constant which depends only on the initial entropy of the particle. This explicit relation between density and pressure along particle paths is sometimes useful. Of course, if the initial entropy varies in space then $\hat{\kappa}$ will be different along different particle paths.

5.2 Isentropic flow

If the entropy is constant everywhere then (5.29) holds with the same value of $\hat{\kappa}$ everywhere and the Euler equations simplify. This is the case, for example, if we consider fluid flow that starts at a uniform rest state (so S is constant) and remains smooth (so S remains constant). Then using (5.29), the equation of state (5.22) reduces to an explicit expression for E in terms of ρ and ρv. The energy equation then becomes redundant and the Euler equations reduce to a system of two equations, the **isentropic equations**,

$$\begin{bmatrix} \rho \\ \rho v \end{bmatrix}_t + \begin{bmatrix} \rho v \\ \rho v^2 + \hat{\kappa}\rho^\gamma \end{bmatrix}_x = 0. \qquad (5.30)$$

5.3   Isothermal flow

Taking $\gamma = 1$ in the isentropic equations (5.30) gives a particularly simple set of equations. This will prove to be a useful example for illustrating the theory presented later since the algebra is relatively simple and yet the behavior is analogous to what is seen for the full Euler equations. By the above discussion, the case $\gamma = 1$ is not physically relevant. However, these same equations can be derived by considering a different physical setup, the case of "isothermal" flow.

Here we consider the flow of gas in a tube that is immersed in a bath at a constant temperature $\bar{T}$, and assume that this bath maintains a constant temperature within the gas. Then the ideal gas law (5.8) reduces to

$$p = a^2 \rho \tag{5.31}$$

where $a^2 \equiv \mathcal{R}\bar{T}$ is a constant and a is the sound speed. Note that maintaining this constant temperature requires heat flux through the wall of the tube, and so energy is no longer conserved in the tube. But mass and momentum are still conserved and these equations, together with the equation of state (5.31), lead to the **isothermal equations**,

$$\begin{bmatrix} \rho \\ \rho v \end{bmatrix}_t + \begin{bmatrix} \rho v \\ \rho v^2 + a^2 \rho \end{bmatrix}_x = 0. \tag{5.32}$$

5.4   The shallow water equations

The study of wave motion in shallow water leads to a system of conservation laws with a similar structure. To derive the one-dimensional equations, we consider fluid in a channel and assume that the vertical velocity of the fluid is negligible and the horizontal velocity $v(x,t)$ is roughly constant through any vertical cross section. This is true if we consider small amplitude waves in a fluid that is shallow relative to the wave length.

We now assume the fluid is incompressible, so the density $\bar{\rho}$ is constant. Instead the height $h(x,t)$ varies, and so the total mass in $[x_1, x_2]$ at time t is

$$\text{total mass in } [x_1, x_2] = \int_{x_1}^{x_2} \bar{\rho} h(x,t) \, dx.$$

The momentum at each point is $\bar{\rho} v(x,t)$ and integrating this vertically gives the mass flux to be $\bar{\rho} v(x,t) h(x,t)$. The constant $\bar{\rho}$ drops out of the conservation of mass equation, which then takes the familiar form

$$h_t + (vh)_x = 0. \tag{5.33}$$

The conservation of momentum equation also takes the same form as in the Euler equations,

$$(\bar{\rho} h v)_t + (\bar{\rho} h v^2 + p)_x = 0, \tag{5.34}$$

but now the pressure p is determined from a hydrostatic law, stating that the pressure at depth y is $\bar{\rho}gy$, where g is the gravitational constant. Integrating this vertically from $y = 0$ to $y = h(x, t)$ gives the total pressure felt at (x, t), the proper pressure term in the momentum flux:

$$p = \frac{1}{2}\bar{\rho}gh^2. \tag{5.35}$$

Using this in (5.34) and cancelling $\bar{\rho}$ gives

$$(hv)_t + \left(hv^2 + \frac{1}{2}gh^2\right)_x = 0. \tag{5.36}$$

Note that the system (5.33), (5.36) is equivalent to the isentropic equation (5.30) in the case $\gamma = 2$.

Equation (5.36) can be simplified by expanding the derivatives and using (5.33) to replace the h_t term. Then several terms drop out and (5.36) is reduced to

$$v_t + \left(\frac{1}{2}v^2 + gh\right)_x = 0. \tag{5.37}$$

Finally, the explicit dependence on g can be eliminated by introducing the variable $\varphi = gh$ into (5.33) and (5.37). The system for the **one-dimensional shallow water equations** then becomes

$$\begin{bmatrix} v \\ \varphi \end{bmatrix}_t + \begin{bmatrix} v^2/2 + \varphi \\ v\varphi \end{bmatrix}_x = 0. \tag{5.38}$$

6 Linear Hyperbolic Systems

In this chapter we begin the study of systems of conservation laws by reviewing the theory of a constant coefficient linear hyperbolic system. Here we can solve the equations explicitly by transforming to characteristic variables. We will also obtain explicit solutions of the Riemann problem and introduce a "phase space" interpretation that will be very useful in our study of nonlinear systems.

Consider the linear system

$$u_t + A u_x = 0 \tag{6.1}$$
$$u(x, 0) = u_0(x)$$

where $u : \mathbb{R} \times \mathbb{R} \to \mathbb{R}^m$ and $A \in \mathbb{R}^{m \times m}$ is a constant matrix. This is a system of conservation laws with the flux function $f(u) = Au$. This system is called **hyperbolic** if A is diagonalizable with real eigenvalues, so that we can decompose

$$A = R\Lambda R^{-1} \tag{6.2}$$

where $\Lambda = \mathrm{diag}(\lambda_1, \lambda_2, \ldots, \lambda_m)$ is a diagonal matrix of eigenvalues and $R = [r_1 | r_2 | \cdots | r_m]$ is the matrix of right eigenvectors. Note that $AR = R\Lambda$, i.e.,

$$A r_p = \lambda_p r_p \quad \text{for } p = 1,\ 2,\ \ldots,\ m. \tag{6.3}$$

The system is called **strictly hyperbolic** if the eigenvalues are distinct. We will always make this assumption as well. (For a demonstration that the behavior can change dramatically in the nonstrictly hyperbolic case, see Exercise 7.4.)

6.1 Characteristic variables

We can solve (6.1) by changing to the "characteristic variables"

$$v = R^{-1} u. \tag{6.4}$$

Multiplying (6.1) by R^{-1} and using (6.2) gives

$$R^{-1}u_t + \Lambda R^{-1}u_x = 0 \qquad (6.5)$$

or, since R^{-1} is constant,

$$v_t + \Lambda v_x = 0. \qquad (6.6)$$

Since Λ is diagonal, this decouples into m independent scalar equations

$$(v_p)_t + \lambda_p(v_p)_x = 0, \qquad p = 1, \ 2, \ \ldots, \ m. \qquad (6.7)$$

Each of these is a constant coefficient linear advection equation, with solution

$$v_p(x,t) = v_p(x - \lambda_p t, 0). \qquad (6.8)$$

Since $v = R^{-1}u$, the initial data for v_p is simply the pth component of the vector

$$v(x,0) = R^{-1}u_0(x). \qquad (6.9)$$

The solution to the original system is finally recovered via (6.4):

$$u(x,t) = Rv(x,t). \qquad (6.10)$$

Note that the value $v_p(x,t)$ is the coefficient of r_p in an eigenvector expansion of the vector $u(x,t)$, i.e., (6.10) can be written out as

$$u(x,t) = \sum_{p=1}^{m} v_p(x,t)r_p. \qquad (6.11)$$

Combining this with the solutions (6.8) of the decoupled scalar equations gives

$$u(x,t) = \sum_{p=1}^{m} v_p(x - \lambda_p t, 0)r_p. \qquad (6.12)$$

Note that $u(x,t)$ depends only on the initial data at the m points $x - \lambda_p t$, so the domain of dependence is given by $\mathcal{D}(\bar{x}, \bar{t}) = \{x = \bar{x} - \lambda_p \bar{t}, \ p = 1, \ 2, \ \ldots, \ m\}$.

The curves $x = x_0 + \lambda_p t$ satisfying $x'(t) = \lambda_p$ are the "characteristics of the pth family", or simply "p-characteristics". These are straight lines in the case of a constant coefficient system. Note that for a strictly hyperbolic system, m distinct characteristic curves pass through each point in the x-t plane. The coefficient $v_p(x,t)$ of the eigenvector r_p in the eigenvector expansion (6.11) of $u(x,t)$ is constant along any p-characteristic.

6.2   Simple waves

We can view the solution as being the superposition of m waves, each of which is advected independently with no change in shape. The pth wave has shape $v_p(x, 0)r_p$ and propagates with speed λ_p. This solution has a particularly simple form if $v_p(x, 0)$ is constant in x for all but one value of p, say $v_p(x, 0) \equiv c_p$ for $p \neq i$. Then the solution has the form

$$u(x, t) = \sum_{p \neq i} c_p r_p + v_i(x - \lambda_i t, 0)r_i \qquad (6.13)$$

$$= u_0(x - \lambda_i t)$$

and the initial data simply propagates with speed λ_i. Since $m - 1$ of the characteristic variables are constant, the equation essentially reduces to $u_t + \lambda_i u_x = 0$ which governs the behavior of the ith family. Nonlinear equations have analogous solutions, called "simple waves", in which variations occur only in one characteristic family (see [97]).

6.3   The wave equation

The canonical example of a hyperbolic PDE is the second order scalar wave equation,

$$u_{tt} = c^2 u_{xx}, \qquad -\infty < x < \infty \qquad (6.14)$$

with initial data

$$u(x, 0) = u_0(x)$$
$$u_t(x, 0) = u_1(x).$$

This can be rewritten as a first order system of conservation laws by introducing

$$v = u_x \quad \text{and} \quad w = u_t$$

Then $v_t = w_x$ by equality of mixed partials, and (6.14) becomes $w_t = c^2 v_x$, so we obtain the system

$$\begin{bmatrix} v \\ w \end{bmatrix}_t + \begin{bmatrix} -w \\ -c^2 v \end{bmatrix}_x = 0, \qquad (6.15)$$

which is of the form (6.1) with

$$A = \begin{bmatrix} 0 & -1 \\ -c^2 & 0 \end{bmatrix}. \qquad (6.16)$$

The initial conditions become

$$v(x, 0) = u_0'(x) \qquad (6.17)$$
$$w(x, 0) = u_1(x).$$

Note that this system is hyperbolic since the eigenvalues of A are $\pm c$, which are the wave speeds.

EXERCISE 6.1. *Compute the eigenvectors of A in (6.16) and use these to decouple (6.15) into a pair of scalar equations. Solve these equations to find that*

$$v(x,t) = \frac{1}{2}\left[u_0'(x_1) + \frac{1}{c}u_1(x_1) + u_0'(x_2) - \frac{1}{c}u_1(x_2)\right] \qquad (6.18)$$

$$w(x,t) = \frac{1}{2}[cu_0'(x_1) + u_1(x_1) - cu_0'(x_2) + u_1(x_2)] \qquad (6.19)$$

where $x_1 = x + ct$, $x_2 = x - ct$. Use this to compute $u(x,t)$ in the original wave equation (6.14).

EXERCISE 6.2. *Rewrite the 2D wave equation*

$$u_{tt} = c^2(u_{xx} + u_{yy}) \qquad (6.20)$$

as a first order system of the form $q_t + Aq_x + Bq_y = 0$ where $q = [v,\ w,\ \varphi]$ by introducing $v = u_x$, $w = u_y$, $\varphi = u_t$.

EXERCISE 6.3. *Solve the "linearized shallow water equations",*

$$\begin{bmatrix} u \\ \varphi \end{bmatrix}_t + \begin{bmatrix} \bar{u} & 1 \\ \bar{\varphi} & \bar{u} \end{bmatrix}\begin{bmatrix} u \\ \varphi \end{bmatrix}_x = 0 \qquad (6.21)$$

where $\bar{u}$ and $\bar{\varphi} > 0$ are constants, with some given initial conditions for u and φ.

Propagation of singularities. If we consider a point $(\bar{x}, \bar{t})$ for which the initial data is smooth at each point $x_0 = \bar{x} - \lambda_p \bar{t}$, $p = 1, 2, \ldots, m$, then $v(x - \lambda_p t, 0) = R^{-1}u(x - \lambda_p t, 0)$ is also a smooth function at each of these points, and hence so is $u(x, t)$. It follows that any singularities in the initial data can propagate only along characteristics, just as in the scalar linear case. Moreover, smooth initial data gives smooth solutions.

6.4 Linearization of nonlinear systems

Now consider a nonlinear system of conservation laws

$$u_t + f(u)_x = 0, \qquad (6.22)$$

where $u : \mathbb{R} \times \mathbb{R} \to \mathbb{R}^m$ and $f : \mathbb{R}^m \to \mathbb{R}^m$. This can be written in the quasilinear form

$$u_t + A(u)u_x = 0 \qquad (6.23)$$

where $A(u) = f'(u)$ is the $m \times m$ Jacobian matrix. Again the system is **hyperbolic** if $A(u)$ is diagonalizable with real eigenvalues for all values of u, at least in some range where the solution is known to lie, and strictly hyperbolic if the eigenvalues are distinct for all u.

We can define characteristics as before by integrating the eigenvalues of $A(u)$. There are m characteristic curves through each point. The curves $x(t)$ in the pth family satisfy

$$x'(t) = \lambda_p(u(x(t), t)),$$
$$x(0) = x_0,$$

for some x_0. Note that the λ_p now depend on u, the solution of the problem, and so we can no longer solve the problem by first determining the characteristics and then solving a system of ODEs along the characteristics. Instead a more complicated coupled system is obtained, reducing the effectiveness of this approach. Characteristics do yield valuable information about what happens locally for smooth data, however. In particular, if we linearize the problem about a constant state $\bar{u}$ we obtain a constant coefficient linear system, with the Jacobian frozen at $A(\bar{u})$. This is relevant if we consider the propagation of small disturbances as we did for the scalar traffic flow model in Chapter 4. Assume an expansion of the solution with the form

$$u(x, t) = \bar{u} + \epsilon u^{(1)}(x, t) + \epsilon^2 u^{(2)}(x, t) + \cdots, \tag{6.24}$$

where $\bar{u}$ is constant and ϵ is small, then by the same derivation as equation (4.12) we find that $u^{(1)}(x, t)$ satisfies

$$u_t^{(1)}(x, t) + A(\bar{u}) u_x^{(1)}(x, t) = 0. \tag{6.25}$$

Small disturbances thus propagate (approximately) along characteristic curves of the form $x_p(t) = x_0 + \lambda_p(\bar{u}) t$. Higher order corrections for nonlinear problems can be obtained by retaining more terms in the expansion. The propagation of discontinuities and other singularities can also be studied through the use of similar expansions. This "geometrical optics" approach to studying weakly nonlinear phenomenon is a powerful tool, but one we will not pursue here. (See Whitham[97], for example.)

Weak shocks. Recall that for a linear system singularities propagate only along characteristics. For nonlinear problems this is not the case, as we have already seen for nonlinear scalar equations in Chapter 3. Instead, the Rankine-Hugoniot jump condition (3.33) must be satisfied for a propagating discontinuity,

$$f(u_l) - f(u_r) = s(u_l - u_r) \tag{6.26}$$

where s is the propagation speed. However, for very weak discontinuities this relates nicely to the linear theory. Suppose that

$$\|u_r - u_l\| \equiv \epsilon \ll 1. \tag{6.27}$$

In this case expanding $f(u_l)$ about u_r gives

$$f(u_l) = f(u_r) + f'(u_r)(u_l - u_r) + O(\epsilon^2) \tag{6.28}$$

so that (6.26) yields

$$f'(u_r)(u_l - u_r) = s(u_l - u_r) + O(\epsilon^2). \tag{6.29}$$

In the limit as $\epsilon \to 0$, the normalized vector $(u_l - u_r)/\epsilon$ must approach an eigenvector of $f'(u_r) = A(u_r)$ with s approaching the corresponding eigenvalue. This observation will be very important in studying the structure of nonlinear solutions more generally.

6.4.1 Sound waves

A familiar example of small disturbances in gas dynamics is the propagation of sound waves in the air. If we consider the Euler equations for a polytropic gas, in which the flux is (2.14) with equation of state (5.22), then the flux Jacobian matrix is

$$f'(u) = \begin{bmatrix} 0 & 1 & 0 \\ \frac{1}{2}(\gamma - 3)v^2 & (3 - \gamma)v & (\gamma - 1) \\ \frac{1}{2}(\gamma - 1)v^3 - v(E + p)/\rho & (E + p)/\rho - (\gamma - 1)v^2 & \gamma v \end{bmatrix}. \tag{6.30}$$

A tedious calculation confirms that the eigenvalues are given by $\lambda_1 = v - c$, $\lambda_2 = v$, and $\lambda_3 = v + c$, where

$$c = \sqrt{\gamma p/\rho}. \tag{6.31}$$

This is the local **speed of sound** in the gas. If we linearize these equations about some state $\bar{u}$, we see that small disturbances propagate at speeds $\bar{v}$, $\bar{v} \pm \bar{c}$, i.e. at speeds 0, $\pm \bar{c}$ relative to the background velocity $\bar{v}$. The waves traveling at speeds $\pm \bar{c}$ are sound waves; our ears are sensitive to the small variations in pressure in these waves.

The waves that travel with the velocity of the gas turn out to be simply density variations which advect with the fluid just as in the scalar linear advection equation. Even for the full nonlinear equations it is easy to check that there are solutions of the form

$$\begin{aligned} \rho(x, t) &= \hat{\rho}(x - \bar{v}t) \\ v(x, t) &= \bar{v} \\ p(x, t) &= \bar{p} \end{aligned} \tag{6.32}$$

where $\hat{\rho}(x)$ is any density distribution.

EXERCISE 6.4. *Verify that (6.32) satisfies the nonlinear Euler equations. Is this solution isentropic?*

EXERCISE 6.5. *The sound speed for the Euler equations is most easily calculated using the equations in the form (5.26), where the entropy S is one of the variables and the equation of state gives $p = p(\rho, S)$. Linearize these equations about a constant state and compute the eigenvalues of the resulting matrix. This verifies that more generally the sound speed is given by $\sqrt{\partial p / \partial \rho}$, which for the polytropic case reduces to (6.31).*

EXERCISE 6.6. *Show that the sound speed in the isentropic equations (5.30) is the same as in the full Euler equations. For this system of two equations, small disturbance waves travel with speeds $\bar{u} \pm \bar{c}$. Waves of the form (6.32) traveling with speed $\bar{u}$ are not isentropic and no longer appear.*

EXERCISE 6.7. *Verify that linearizing the shallow water equations (5.38) gives the system (6.21). What is the "sound speed" for this system?*

6.5   The Riemann Problem

For the constant coefficient linear system, the Riemann problem can be explicitly solved. We will see shortly that the solution to a nonlinear Riemann problem has a simple structure which is quite similar to the structure of this linear solution, and so it is worthwhile studying the linear case in some detail.

The Riemann problem consists of the equation $u_t + A u_x = 0$ together with piecewise constant initial data of the form

$$u(x, 0) = \begin{cases} u_l & x < 0 \\ u_r & x > 0 \end{cases} \tag{6.33}$$

We will assume that this system is **strictly hyperbolic**, which means that the eigenvalues of A are real and distinct, and order them

$$\lambda_1 < \lambda_2 < \cdots < \lambda_m. \tag{6.34}$$

Recall that the general solution to the linear problem is given by (6.12). For the Riemann problem we can simplify the notation if we decompose u_l and u_r as

$$u_l = \sum_{p=1}^{m} \alpha_p r_p \qquad u_r = \sum_{p=1}^{m} \beta_p r_p. \tag{6.35}$$

Then

$$v_p(x, 0) = \begin{cases} \alpha_p & x < 0 \\ \beta_p & x > 0 \end{cases} \tag{6.36}$$

and so

$$v_p(x, t) = \begin{cases} \alpha_p & \text{if } x - \lambda_p t < 0 \\ \beta_p & \text{if } x - \lambda_p t > 0. \end{cases} \tag{6.37}$$

If we let $P(x,t)$ be the maximum value of p for which $x - \lambda_p t > 0$, then

$$u(x,t) = \sum_{p=1}^{P(x,t)} \beta_p r_p + \sum_{p=P(x,t)+1}^{m} \alpha_p r_p. \tag{6.38}$$

The determination of $u(x,t)$ at a given point is illustrated in Figure 6.1. In the case shown, $v_1 = \beta_1$ while $v_2 = \alpha_2$ and $v_3 = \alpha_3$. The solution at the point illustrated is thus

$$u(x,t) = \beta_1 r_1 + \alpha_2 r_2 + \alpha_3 r_3. \tag{6.39}$$

Note that the solution is the same at any point in the wedge between the $x' = \lambda_1$ and $x' = \lambda_2$ characteristics. As we cross the pth characteristic, the value of $x - \lambda_p t$ passes through 0 and the corresponding v_p jumps from α_p to β_p. The other coefficients v_i $(i \neq p)$ remain constant.

The solution is constant in each of the wedges as shown in Figure 6.2. Across the pth characteristic the solution jumps with the jump given by

$$[u] = (\beta_p - \alpha_p)r_p. \tag{6.40}$$

Note that these jumps satisfy the Rankine-Hugoniot conditions (3.33), since $f(u) = Au$ leads to

$$\begin{aligned} [f] &= A[u] \\ &= (\beta_p - \alpha_p)Ar_p \\ &= \lambda_p[u] \end{aligned}$$

and λ_p is precisely the speed of propagation of this jump. The solution $u(x,t)$ in (6.38) can alternatively be written in terms of these jumps as

$$\begin{aligned} u(x,t) &= u_l + \sum_{\lambda_p < x/t} (\beta_p - \alpha_p)r_p \tag{6.41} \\ &= u_r - \sum_{\lambda_p \geq x/t} (\beta_p - \alpha_p)r_p \tag{6.42} \end{aligned}$$

It might happen that the initial jump $u_r - u_l$ is already an eigenvector of A, if $u_r - u_l = (\beta_i - \alpha_i)r_i$ for some i. In this case $\alpha_p = \beta_p$ for $p \neq i$. Then this discontinuity simply propagates with speed λ_i, and the other characteristics carry jumps of zero strength.

In general this is not the case, however, and the jump $u_r - u_l$ cannot propagate as a single discontinuity with any speed without violating the Rankine-Hugoniot condition. We can view "solving the Riemann problem" as finding a way to split up the jump $u_r - u_l$ into a sum of jumps

$$u_r - u_l = (\beta_1 - \alpha_1)r_1 + \cdots + (\beta_m - \alpha_m)r_m, \tag{6.43}$$

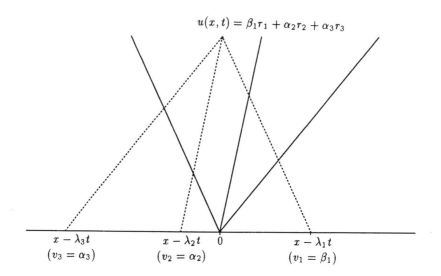

Figure 6.1. Construction of solution to Riemann problem at (x, t).

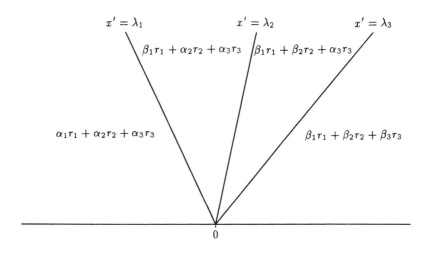

Figure 6.2. Values of solution u in each wedge of x–t plane.

each of which *can* propagate at an appropriate speed λ_i with the Rankine-Hugoniot condition satisfied.

For nonlinear systems we solve the Riemann problem in much the same way: The jump $u_r - u_l$ will usually not have the property that $[f]$ is a scalar multiple of $[u]$, but we can attempt to find a way to split this jump up as a sum of jumps, across each of which this property does hold. (Although life is complicated by the fact that we may need to introduce rarefaction waves as well as shocks.) In studying the solution of the Riemann problem, the jump in the pth family, traveling at constant speed λ_p, is often called the p-wave.

6.5.1 The phase plane

For systems of two equations, it is illuminating to view this splitting in the phase plane. This is simply the u_1-u_2 plane, where $u = (u_1, u_2)$. Each vector $u(x, t)$ is represented by a point in this plane. In particular, u_l and u_r are points in this plane and a discontinuity with left and right states u_l and u_r can propagate as a single discontinuity only if $u_r - u_l$ is an eigenvector of A, which means that the line segment from u_l to u_r must be parallel to the eigenvector r_1 or r_2. Figure 6.3 shows an example. For the state u_l illustrated there, the jump from u_l to u_r can propagate as a single discontinuity if and only if u_r lies on one of the two lines drawn through u_l in the direction r_1 and r_2. These lines give the locus of all points that can be connected to u_l by a 1-wave or a 2-wave. This set of states is called the **Hugoniot locus**. We will see that there is a direct generalization of this to nonlinear systems in the next chapter.

Similarly, there is a Hugoniot locus through any point u_r that gives the set of all points u_l that can be connected to u_r by an elementary p-wave. These curves are again in the directions r_1 and r_2.

For a general Riemann problem with arbitrary u_l and u_r, the solution consists of two discontinuities travelling with speeds λ_1 and λ_2, with a new constant state in between that we will call u_m. By the discussion above,

$$u_m = \beta_1 r_1 + \alpha_2 r_2 \tag{6.44}$$

so that $u_m - u_l = (\beta_1 - \alpha_1)r_1$ and $u_r - u_m = (\beta_2 - \alpha_2)r_2$. The location of u_m in the phase plane must be where the 1-wave locus through u_l intersects the 2-wave locus through u_r. This is illustrated in Figure 6.4a.

Note that if we interchange u_r and u_l in this picture, the location of u_m changes as illustrated in Figure 6.4b. In each case we travel from u_l to u_r by first going in the direction r_1 and then in the direction r_2. This is required by the fact that $\lambda_1 < \lambda_2$ since clearly the jump between u_l and u_m must travel slower than the jump between u_m and u_r if we are to obtain a single-valued solution.

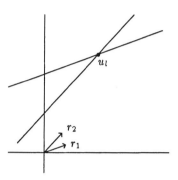

Figure 6.3. The Hugoniot locus of the state u_l consists of all states that differ from u_l by a scalar multiple of r_1 or r_2.

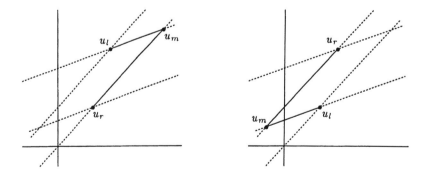

Figure 6.4. The new state u_m arising in the solution to the Riemann problem for two different choices of u_l and u_r.

For systems with more than two equations, the same interpretation is possible but becomes harder to draw since the phase space is now m dimensional. Since the m eigenvectors r_p are linearly independent, we can decompose any jump $u_r - u_l$ into the sum of jumps in these directions, obtaining a piecewise linear path from u_l to u_r in m-dimensional space.

7 Shocks and the Hugoniot Locus

We now return to the nonlinear system $u_t + f(u)_x = 0$, where $u(x,t) \in \mathbb{R}^m$. As before we assume strict hyperbolicity, so that $f'(u)$ has disctinct real eigenvalues $\lambda_1(u) < \cdots < \lambda_m(u)$ and hence linearly independent eigenvectors. We choose a particular basis for these eigenvectors, $\{r_p(u)\}_{p=1}^m$, usually chosen to be normalized in some manner, e.g. $\|r_p(u)\| \equiv 1$.

In the previous chapter we constructed the solution to the general Riemann problem for a linear hyperbolic system of conservation laws. Our goal in the next two chapters is to perform a similar construction for the nonlinear Riemann problem. In the linear case the solution consists of m waves, which are simply discontinuities traveling at the characteristic velocities of the linear system. In the nonlinear case our experience with the scalar equation leads us to expect more possibilities. In particular, the physically relevant vanishing viscosity solution may contain rarefaction waves as well as discontinuities. In this chapter we will first ignore the entropy condition and ask a simpler question: is it possible to construct a weak solution of the Riemann problem consisting only of m discontinuities propagating with constant speeds $s_1 < s_2 < \cdots < s_m$? As we will see, the answer is yes for $\|u_l - u_r\|$ sufficiently small.

7.1 The Hugoniot locus

Recall that if a discontinuity propagating with speed s has constant values $\hat{u}$ and $\tilde{u}$ on either side of the discontinuity, then the Rankine-Hugoniot jump condition must hold,

$$f(\tilde{u}) - f(\hat{u}) = s(\tilde{u} - \hat{u}). \tag{7.1}$$

Now suppose we fix the point $\hat{u} \in \mathbb{R}^m$ and attempt to determine the set of all points $\tilde{u}$ which can be connected to $\hat{u}$ by a discontinuity satisfying (7.1) for some s. This gives a system of m equations in $m+1$ unknowns: the m components of $\tilde{u}$, and s. This leads us to expect one parameter families of solutions.

We know that in the linear case there are indeed m such families for any $\hat{u}$. In the pth family the jump $\tilde{u} - \hat{u}$ is some scalar multiple of r_p, the pth eigenvector of A. We can parameterize these families of solutions using this scalar multiple, say ξ, and we obtain the following solution curves:

$$\tilde{u}_p(\xi; \hat{u}) = \hat{u} + \xi r_p$$
$$s_p(\xi; \hat{u}) = \lambda_p$$

for $p = 1, 2, \ldots, m$. Note that $\tilde{u}_p(0; \hat{u}) = \hat{u}$ for each p and so through the point $\hat{u}$ in phase space there are m curves (straight lines in fact) of possible solutions. This is illustrated in Figure 6.4 for the case $m = 2$. The two lines through each point are the states that can be connected by a discontinuity with jump proportional to r_1 or r_2.

In the nonlinear case we also obtain m curves through any point $\hat{u}$, one for each characteristic family. We again parameterize these curves by $\tilde{u}_p(\xi; \hat{u})$ with $\tilde{u}_p(0; \hat{u}) = \hat{u}$ and let $s_p(\xi; \hat{u})$ denote the corresponding speed. To simplify notation, we will frequently write these as simply $\tilde{u}_p(\xi)$, $s_p(\xi)$ when the point $\hat{u}$ is clearly understood.

The Rankine-Hugoniot condition gives

$$f(\tilde{u}_p(\xi)) - f(\hat{u}) = s_p(\xi)(\tilde{u}_p(\xi) - \hat{u}). \tag{7.2}$$

Differentiating this expression with respect to ξ and setting $\xi = 0$ gives

$$f'(\hat{u})\tilde{u}_p'(0) = s_p(0)\tilde{u}_p'(0) \tag{7.3}$$

so that $\tilde{u}_p'(0)$ must be a scalar multiple of the eigenvector $r_p(\hat{u})$ of $f'(\hat{u})$, while $s_p(0) = \lambda_p(\hat{u})$. The curve $\tilde{u}_p(\xi)$ is thus tangent to $r_p(\hat{u})$ at the point $\hat{u}$. We have already observed this, by a slightly different argument, in discussing weak shocks in Chapter 6. For a system of $m = 2$ equations, this is easily illustrated in the phase plane. An example for the isothermal equations of gas dynamics is discussed below, see Figure 7.1.

For smooth f, it can be shown using the implicit function theorem that these solution curves exist locally in a neighborhood of $\hat{u}$, and that the functions $\tilde{u}_p$ and s_p are smooth. See Lax[45] or Smoller[77] for details. These curves are called Hugoniot curves. The set of all points on these curves is often collectively called the **Hugoniot locus** for the point $\hat{u}$. If $\tilde{u}_p$ lies on the pth Hugoniot curve through $\hat{u}$, then we say that $\hat{u}$ and $\tilde{u}_p$ are connected by a p-shock.

EXAMPLE 7.1. The isothermal equations of gas dynamics (5.32) provide a relatively simple example of the nonlinear theory.

If we let m represent the momentum, $m = \rho v$, then the system becomes

$$\rho_t + m_x = 0 \tag{7.4}$$
$$m_t + (m^2/\rho + a^2\rho)_x = 0$$

or $u_t + f(u)_x = 0$ where $u = (\rho, m)$.

The Jacobian matrix is

$$f'(u) = \begin{bmatrix} 0 & 1 \\ a^2 - m^2/\rho^2 & 2m/\rho \end{bmatrix}. \tag{7.5}$$

The eigenvalues are

$$\lambda_1(u) = m/\rho - a, \qquad \lambda_2(u) = m/\rho + a \tag{7.6}$$

with eigenvectors

$$r_1(u) = \begin{bmatrix} 1 \\ m/\rho - a \end{bmatrix}, \qquad r_2(u) = \begin{bmatrix} 1 \\ m/\rho + a \end{bmatrix}. \tag{7.7}$$

These could be normalized but it is easiest to leave them in this simple form.

Now let's fix a state $\hat{u}$ and determine the set of states $\tilde{u}$ that can be connected by a discontinuity. The Rankine-Hugoniot condition (7.1) becomes, for this system,

$$\tilde{m} - \hat{m} = s(\tilde{\rho} - \hat{\rho}) \tag{7.8}$$
$$(\tilde{m}^2/\tilde{\rho} + a^2\tilde{\rho}) - (\hat{m}^2/\hat{\rho} + a^2\hat{\rho}) = s(\tilde{m} - \hat{m}).$$

This gives two equations in the three unknowns $\tilde{\rho}$, $\tilde{m}$, and s. These equations can be easily solved for $\tilde{m}$ and s in terms of $\tilde{\rho}$, giving

$$\tilde{m} = \tilde{\rho}\hat{m}/\hat{\rho} \pm a\sqrt{\tilde{\rho}/\hat{\rho}} \, (\tilde{\rho} - \hat{\rho}) \tag{7.9}$$

and

$$s = \hat{m}/\hat{\rho} \pm a\sqrt{\tilde{\rho}/\hat{\rho}}. \tag{7.10}$$

The $\pm$ signs in these equations give two solutions, one for each family. Since $\tilde{m}$ and s can be expressed in terms of $\tilde{\rho}$, we can easily parameterize these curves by taking, for example,

$$\tilde{\rho}_p(\xi; \hat{u}) = \hat{\rho}(1 + \xi), \qquad p = 1, \, 2. \tag{7.11}$$

We then have

$$\tilde{u}_1(\xi; \hat{u}) = \hat{u} + \xi \begin{bmatrix} \hat{\rho} \\ \hat{m} - a\hat{\rho}\sqrt{1 + \xi} \end{bmatrix}, \qquad s_1(\xi; \hat{u}) = \hat{m}/\hat{\rho} - a\sqrt{1 + \xi}. \tag{7.12}$$

and

$$\tilde{u}_2(\xi; \hat{u}) = \hat{u} + \xi \begin{bmatrix} \hat{\rho} \\ \hat{m} + a\hat{\rho}\sqrt{1 + \xi} \end{bmatrix}, \qquad s_2(\xi; \hat{u}) = \hat{m}/\hat{\rho} + a\sqrt{1 + \xi}. \tag{7.13}$$

The choice of signs for each family is determined by the behavior as $\xi \to 0$, where the relation (7.3) must hold. It is easy to check that with the above choice we have

$$\frac{\partial}{\partial \xi}\tilde{u}_p(0; \hat{u}) = \hat{\rho} r_p(\hat{u}) \propto r_p(\hat{u}),$$
$$s_p(0; \hat{u}) = \lambda_p(\hat{u}),$$

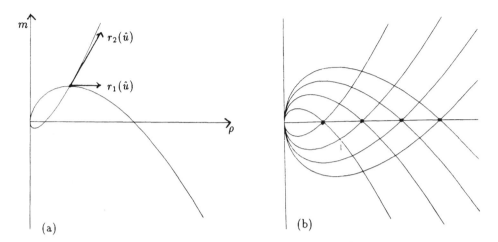

Figure 7.1. a) Hugoniot locus for the state $\hat{u} = (1,1)$ in the isothermal gas dynamics equations with $a = 1$. b) Variation of these curves for $\hat{u} = (\hat{\rho}, 0)$, $\hat{\rho} = 1, 2, 3, 4$.

as expected.

Notice that real-valued solutions exist only for $\xi > -1$ and that $\tilde{u}_p(-1; \hat{u}) = (0,0)$ for $p = 1, 2$ and any $\hat{u}$. Thus, each Hugoniot locus terminates at the origin (the *vacuum state*, since $\rho = 0$). There are no states with $\rho < 0$ that can be connected to $\hat{u}$ by a propagating discontinuity. The curves $\tilde{u}_p(\xi)$ are illustrated in Figure 7.1a for one particular choice of $\hat{u}$ and $a = 1$. Figure 7.1b shows how these curves vary with $\hat{u}$. In this case we see $\tilde{u}_p(\xi; \hat{u})$ for $\hat{u} = (\hat{\rho}, 0)$, $\hat{\rho} = 1, 2, 3, 4$.

EXERCISE 7.1. *Determine the Hugoniot locus for the shallow water equations (5.38).*

7.2   Solution of the Riemann problem

Now suppose that we wish to solve the Riemann problem with left and right states u_l and u_r (and recall that we are ignoring the entropy condition at this point). Just as in the linear case, we can accomplish this by finding an intermediate state u_m such that u_l and u_m are connected by a discontinuity satisfying the Rankine-Hugoniot condition, and so are u_m and u_r. Graphically we accomplish this by drawing the Hugoniot locus for each of the states u_l and u_r and looking for intersections. See Figure 7.2 for an example with the isothermal equations.

In this example there are two points of intersection, labelled u_m and u_m^*, but only u_m gives a single-valued solution to the Riemann problem since we need the jump from

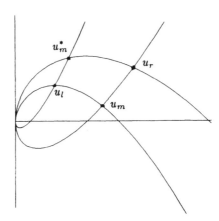

Figure 7.2. Construction of a weak solution to the Riemann problem with left and right states u_l and u_r.

u_l to u_m to travel more slowly than the jump from u_m to u_r. This requires that u_m be connected to u_l by a 1-shock while u_r is connected to u_m by a 2-shock, due to our convention that $\lambda_1(u) < \lambda_2(u)$. Of course our requirement really involves the shock speeds, not the characteristic speeds, but note that any 1-shock connected to u_m has speed

$$s_1(\xi; u_m) = m_m/\rho_m - a\sqrt{1 + \xi/\rho_m} < m_m/\rho_m \qquad \forall \xi > -\rho_m$$

while any 2-shock has speed

$$s_2(\xi; u_m) = m_m/\rho_m + a\sqrt{1 + \xi/\rho_m} > m_m/\rho_m \qquad \forall \xi > -\rho_m$$

and consequently $s_1(\xi_1; u_m) < s_2(\xi_2; u_m)$ for all ξ_1, $\xi_2 > -\rho_m$.

The state u_m can be found algebraically by using our explicit expressions for the Hugoniot locus. We wish to find a state (ρ_m, m_m) which is connected to u_l by a 1-shock and to u_r by a 2-shock. Consequently equation (7.9) with the minus sign should hold for $\tilde{u} = u_m$, $\hat{u} = u_l$, and the same equation with the plus sign should hold for $\tilde{u} = u_m$, $\hat{u} = u_r$. Equating the two resulting expressions for m_m gives

$$\rho_m m_l/\rho_l - a\sqrt{\rho_m/\rho_l}\,(\rho_m - \rho_l) = \rho_m m_r/\rho_r + a\sqrt{\rho_m/\rho_r}\,(\rho_m - \rho_r). \qquad (7.14)$$

Setting $z = \sqrt{\rho_m}$, we obtain a quadratic equation for z,

$$\left(\frac{a}{\sqrt{\rho_r}} + \frac{a}{\sqrt{\rho_l}}\right) z^2 + \left(\frac{m_r}{\rho_r} - \frac{m_l}{\rho_l}\right) z - a\left(\sqrt{\rho_r} + \sqrt{\rho_l}\right) = 0. \qquad (7.15)$$

This equation has a unique positive solution $\bar{z}$, and $\rho_m = \bar{z}^2$. We can then determine m_m by evaluating either side of (7.14).

More generally, for a system of m equations we can attempt to solve the Riemann problem by finding a sequence of states u_1, u_2, ..., u_{m-1} such that u_l is connected to u_1 by a 1-shock, u_1 is connected to u_2 by a 2-shock, and so on, with u_{m-1} connected to u_r by an m-shock. If u_l and u_r are sufficiently close together then this can always be achieved. Lax proved a stronger version of this in his fundamental paper [44]. (By considering rarefaction waves also, the entropy satisfying solution can be constructed in a similar manner.) See also Lax[45] and Smoller[77] for detailed proofs. The idea is quite simple and will be summarized here following Lax[45]. From u_l we know that we can reach a one parameter family of states $u_1(\xi_1)$ through a 1-shock. From $u_1(\xi_1)$ we can reach another one parameter family of states $u_2(\xi_1, \xi_2)$ through a 2-shock. Continuing, we see that from u_l we can reach an m parameter family of states $u_m(\xi_1, \xi_2, \dots, \xi_m)$. Moreover, we know that

$$\frac{\partial u_m}{\partial \xi_p}\bigg|_{\xi_1 = \dots = \xi_m = 0} \propto r_p(u_l), \qquad p = 1, 2, \dots, m.$$

These vectors are linearly independent by our hyperbolicity assumption, and hence the Jacobian of the mapping $u_m : \mathbb{R}^m \to \mathbb{R}^m$ is nonsingular. It follows that the mapping u_m is bijective in a neighborhood of the origin. Hence for any u_r sufficiently close to u_l there is a unique set of parameters ξ_1, ..., ξ_m such that $u_r = u_m(\xi_1, \dots, \xi_m)$.

7.2.1 Riemann problems with no solution

For a general nonlinear system the local result quoted above is the best we can expect, and there are examples where the Riemann problem has no weak solution when $\|u_l - u_r\|$ is sufficiently large. In such examples the shock curves simply do not intersect in the required way. One such example is given by Smoller[76].

For specific systems of conservation laws one can often prove that solutions to the Riemann problem exist more globally. For example, in the isothermal case we can clearly always solve the Riemann problem provided ρ_l, $\rho_r > 0$. The case where ρ_l or ρ_r is negative is of little interest physically, but even then the Riemann problem can be solved by taking the vacuum state $u_m = (0, 0)$ as the intermediate state. We encounter problems only if $\rho = 0$ and $m \neq 0$ for one of the states, but since the equations involve m/ρ this is not surprising.

7.3 Genuine nonlinearity

In defining the Hugoniot locus above, we ignored the question of whether a given discontinuity is physically relevant. The state $\tilde{u}$ is in the Hugoniot locus of $\hat{u}$ if the jump satisfies the Rankine-Hugoniot condition, regardless of whether this jump could exist in a vanishing viscosity solution. We would now like to define an entropy condition that can be applied directly to a discontinuous weak solution to determine whether the jumps should be allowed. In Chapter 3 we considered several forms of the entropy condition for scalar equations. In the convex case, the simplest condition is simply that characteristics should go *into* the shock as time advances, resulting in the requirement (3.45),

$$f'(u_l) > s > f'(u_r).$$

Lax[44] proposed a simple generalization of this entropy condition to systems of equations that are genuinely nonlinear, a natural generalization of the convex scalar equation. The pth characteristic field is said to be **genuinely nonlinear** if

$$\nabla \lambda_p(u) \cdot r_p(u) \neq 0 \qquad \text{for all } u, \tag{7.16}$$

where $\nabla \lambda_p(u) = (\partial \lambda_p / \partial u_1, \ldots, \partial \lambda_p / \partial u_m)$ is the gradient of $\lambda_p(u)$. Note that in the scalar case, $m = 1$ and $\lambda_1(u) = f'(u)$ while $r_1(u) = 1$ for all u. The condition (7.16) reduces to the convexity requirement $f''(u) \neq 0 \ \forall u$. This implies that the characteristic speed $f'(u)$ is monotonically increasing or decreasing as u varies, and leads to a relatively simple solution of the Riemann problem.

For a system of equations, (7.16) implies that $\lambda_p(u)$ is monotonically increasing or decreasing as u varies along an integral curve of the vector field $r_p(u)$. These integral curves will be discussed in detail in the next chapter, where we will see that through a rarefaction wave u varies along an integral curve. Since monotonicity of the propagation speed λ_p is clearly required through a rarefaction wave, genuine nonlinearity is a natural assumption.

7.4   The Lax entropy condition

For a genuinely nonlinear field, Lax's entropy condition says that a jump in the pth field (from u_l to u_r, say) is admissible only if

$$\lambda_p(u_l) > s > \lambda_p(u_r) \tag{7.17}$$

where s is again the shock speed. Characteristics in the pth family disappear into the shock as time advances, just as in the scalar case.

EXAMPLE 7.2.   For isothermal gas dynamics we can easily verify that both fields are genuinely nonlinear. Since $\lambda_p = m/\rho \pm a$, we compute that in each case

$$\nabla \lambda_p(u) = \begin{bmatrix} -m/\rho^2 \\ 1/\rho \end{bmatrix}, \quad p = 1, \ 2. \tag{7.18}$$

Using (7.7), we compute that

$$\nabla\lambda_1(u) \cdot r_1(u) = -a/\rho \tag{7.19}$$

$$\nabla\lambda_2(u) \cdot r_2(u) = a/\rho. \tag{7.20}$$

These quantities are both nonzero for all u.

Now suppose u_l and u_r are connected by a 1-shock. Then u_l lies in the Hugoniot locus of u_r and also u_r lies in the Hugoniot locus of u_l. We can thus evaluate the shock speed s using (7.10) (with the minus sign, since the jump is a 1-shock) in two different ways, obtaining

$$s = \frac{m_l}{\rho_l} - a\sqrt{\rho_r/\rho_l} = \frac{m_r}{\rho_r} - a\sqrt{\rho_l/\rho_r}. \tag{7.21}$$

Since $\lambda_1(u) = m/\rho - a$, the entropy condition (7.17) becomes

$$\frac{m_l}{\rho_l} - a > \frac{m_l}{\rho_l} - a\sqrt{\rho_r/\rho_l} = \frac{m_r}{\rho_r} - a\sqrt{\rho_l/\rho_r} > \frac{m_r}{\rho_r} - a \tag{7.22}$$

and is clearly satisfied if and only if $\rho_r > \rho_l$.

Notice that since the fluid velocity is $v = m/\rho$, 1-shocks always travel slower than the fluid on either side, and so a given fluid particle passes through the shock from left to right (i.e. its state jumps from u_l to u_r). A consequence of the entropy condition is that the density of the gas must *increase* as it goes through the shock. This is also true more generally in the full Euler equations. The gas can only be compressed as the shock passes, not rarefied (rarefaction occurs, naturally enough, through a rarefaction wave rather than a shock).

For 2-shocks the entropy condition requires

$$\frac{m_l}{\rho_l} + a > \frac{m_l}{\rho_l} + a\sqrt{\rho_r/\rho_l} = \frac{m_r}{\rho_r} + a\sqrt{\rho_r/\rho_l} > \frac{m_r}{\rho_r} + a \tag{7.23}$$

which is now satisfied only if $\rho_r < \rho_l$. But note that 2-shocks travel faster than the fluid on either side, so that particles pass through the shock from right to left. So the entropy condition has the same physical interpretation as before: the density must jump from ρ_r to a higher value ρ_l as the gas goes through the shock.

We can now reconsider the Hugoniot locus of a point $\hat{u}$ and retain only the points $\tilde{u}$ that can be connected to $\hat{u}$ by an entropy-satisfying shock, discarding the entropy-violating shocks. In order to do this, we must first decide whether $\hat{u}$ is to lie to the left of the discontinuity or to the right. The entropy condition (7.17), unlike the Rankine-Hugoniot condition, is not symmetric in the two states.

Figure 7.3a shows the set of states that can be connected to the right of a given state $\hat{u}$ by an entropy-satisfying shock. Figure 7.3b shows the set of states that can be connected to the left of the same state $\hat{u}$. Note that the union of these curves gives the full Hugoniot locus. Each branch of the Hugoniot locus splits into two parts at $\hat{u}$; states on one side can only be connected to the left, states on the other side can only be connected to the right.

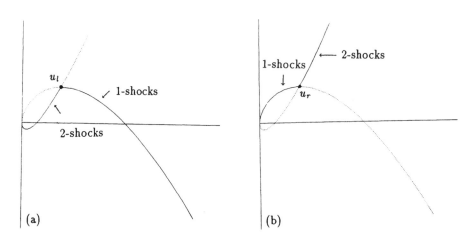

Figure 7.3. a) States u_r that can be connected to $u_l = \hat{u}$ by an entropy-satisfying shock. b) States u_l that can be connected to $u_r = \hat{u}$ by an entropy-satisfying shock. In each case the entropy-violating branches of the Hugoniot locus are shown as dashed lines.

7.5 Linear degeneracy

The assumption of genuine nonlinearity is obviously violated for a constant coefficient linear system, in which $\lambda_p(u)$ is constant and hence $\nabla\lambda_p \equiv 0$. More generally, for a nonlinear system it might happen that in one of the characteristic fields the eigenvalue $\lambda_p(u)$ is constant along integral curves of this field, and hence

$$\nabla\lambda_p(u) \cdot r_p(u) \equiv 0 \qquad \forall u. \tag{7.24}$$

(Of course the value of $\lambda_p(u)$ might vary from one integral curve to the next.) In this case we say that the pth field is **linearly degenerate**. This may seem rather unlikely, and not worth endowing with a special name, but in fact the Euler equations have this property. We will see later that for this system of three equations, two of the fields are genuinely nonlinear while the third is linearly degenerate.

  A discontinuity in a linearly degenerate field is called a **contact discontinuity**. This name again comes from gas dynamics. In a shock tube problem the gas initially on one side of the diaphram never mixes with gas from the other side (in the inviscid Euler equations). As time evolves these two gases remain in contact along a ray in the x-t plane along which, in general, there is a jump in density. This is the contact discontinuity.

  For general systems, if the pth field is linearly degenerate and u_l and u_r are connected by a discontinuity in this field, then it can be shown that u_l and u_r lie on the same integral curve of $r_p(u)$, so that $\lambda_p(u_l) = \lambda_p(u_r)$. Moreover, the propagation speed s is also equal to

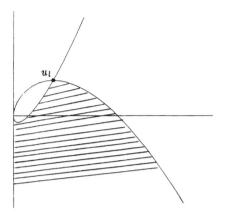

Figure 7.4. The shaded region shows the set of points u_r for which the Riemann problem with states u_l, u_r will have an entropy satisfying solution consisting of two shocks.

$\lambda_p(u_l)$. Consequently, the p-characteristics are parallel to the propagating discontinuity on each side, just as in the case of a linear system. Contact discontinuities can occur in vanishing viscosity solutions, and so for systems with both genuinely nonlinear and linearly degenerate fields, we should modify the entropy condition (7.17) to read

$$\lambda_p(u_l) \geq s \geq \lambda_p(u_r). \tag{7.25}$$

7.6 The Riemann problem

Returning to the Riemann problem, we see that for general data u_l and u_r, the weak solution previously constructed will not be the physically correct solution if any of the resulting shocks violate the entropy condition. In this case we need to also use rarefaction waves in our construction of the solution. This will be done in the next chapter.

EXAMPLE 7.3. In particular, the weak solution to the isothermal equations constructed in Figure 7.2 is nonphysical, since the 2-shock does not satisfy the entropy condition. For the state u_l illustrated there, a physically correct Riemann solution consisting of two shocks will exist only if the state u_r lies in the region shaded in Figure 7.4.

EXERCISE 7.2. *Consider the shallow water equations (5.38).*

1. *Show that a weak solution to the Riemann problem consisting only of shocks always exists if φ_l, $\varphi_r > 0$. Determine the intermediate state u_m for given states u_l and u_r.*

2. Show that both fields are genuinely nonlinear.

3. Give a physical interpretation to the entropy condition for this system.

EXERCISE 7.3. *Consider the system*

$$v_t + [vg(v, \phi)]_x = 0 \tag{7.26}$$
$$\phi_t + [\phi g(v, \phi)]_x = 0$$

where $g(v, \phi)$ is a given function. Systems of this form arise in two-phase flow. As a simple example, take $g(v, \phi) = \phi^2$ and assume $\phi > 0$.

1. Determine the eigenvalues and eigenvectors for this system and show that the first field is linearly degenerate while the second field is genuinely nonlinear.

2. Show that the Hugoniot locus of any point $\hat{u}$ consists of a pair of straight lines, and that each line is also the integral curve of the corresponding eigenvector.

3. Obtain the general solution to the Riemann problem consisting of one shock and one contact discontinuity. Show that this solution satisfies the entropy condition if and only if $\phi_l \geq \phi_r$.

EXERCISE 7.4 (A NONSTRICTLY HYPERBOLIC SYSTEM). *Consider the system*

$$\rho_t + (\rho v)_x = 0 \tag{7.27}$$
$$v_t + \left(\frac{1}{2}v^2\right)_x = 0 \tag{7.28}$$

1. Show that the Jacobian matrix has eigenvalues $\lambda_1 = \lambda_2 = v$ and a one-dimensional space of eigenvectors proportional to $r = (1, v)$. Hence the Jacobian matrix is not diagonalizable, and the system is hyperbolic but not strictly hyperbolic.

2. Note that equation (7.28) decouples from (7.27) and is simply Burgers' equation for v. We can compute $v(x, t)$ from this equation, independent of ρ, and then (7.27) becomes

$$\rho_t + v\rho_x = -\rho v_x$$

where the right hand side is now a source term and v is known. This is an evolution equation for ρ along characteristics. What happens when v becomes discontinuous?

3. The system (7.27), (7.28) is the isothermal system (7.4) in the case $a = 0$. Use the theory of this chapter to investigate the limit $a \to 0$, and determine how the solution to a Riemann problem behaves in this limit. Relate this to the results of Part 2 above.

8 Rarefaction Waves and Integral Curves

All of the Riemann solutions considered so far have the following property: the solution is constant along all rays of the form $x = \xi t$. Consequently, the solution is a function of x/t alone, and is said to be a "similarity solution" of the PDE. A rarefaction wave solution to a system of equations also has this property and takes the form

$$u(x,t) = \begin{cases} u_l & x \leq \xi_1 t \\ w(x/t) & \xi_1 t < x < \xi_2 t \\ u_r & x \geq \xi_2 t \end{cases} \tag{8.1}$$

where w is a smooth function with $w(\xi_1) = u_l$ and $w(\xi_2) = u_r$.

When does a system of equations have a solution of this form? As in the case of shocks, for arbitrary states u_l and u_r there might not be a solution of this form. But in general, starting at each point u_l there are m curves consisting of points u_r which can be connected to u_l by a rarefaction wave. These turn out to be subsets of the integral curves of the vector fields $r_p(u)$.

8.1 Integral curves

An integral curve for $r_p(u)$ is a curve which has the property that the tangent to the curve at any point u lies in the direction $r_p(u)$. The existence of smooth curves of this form follows from smoothness of f and strict hyperbolicity, since $r_p(u)$ is then a smooth function of u. If $u_p(\xi)$ is a parameterization (for $\xi \in \mathbb{R}$) of an integral curve in the pth family, then the tangent vector is proportional to $r_p(u_p(\xi))$ at each point, i.e.

$$u_p'(\xi) = \alpha(\xi) r_p(u_p(\xi)) \tag{8.2}$$

where $\alpha(\xi)$ is some scalar factor.

EXAMPLE 8.1. For the isothermal equations we can easily sketch the integral curves of r_p in the phase plane by drawing a selection of eigenvectors r_p and then sketching in

81

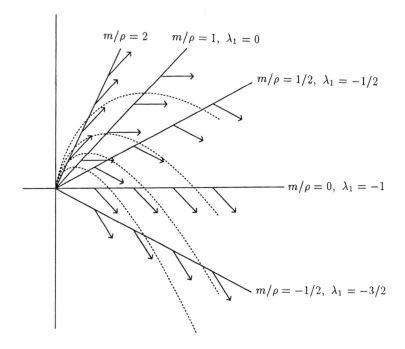

Figure 8.1. Integral curves of r_1 in the phase plane.

the smooth integral curves of this vector field. Figure 8.1 shows an example for the first characteristic field, in the case where the sound speed is taken as $a = 1$. Notice that along any ray $m/\rho = c$ in the phase plane the vector $r_1(u)$ is constant, since by (7.7), $r_1(u) = (1, c - a)$. This observaion has been used in Figure 8.1 to sketch in a number of eigenvectors.

Also notice that along $m/\rho = c$ we have $\lambda_1(u) = c - a$, so λ_1 is also constant along these rays. These are level lines of the function $\lambda_1(u)$. Since this characteristic field is genuinely nonlinear by (7.19), the function $\lambda_1(u)$ varies monotonically along any integral curve of r_1, as is clearly seen in Figure 8.1.

8.2   Rarefaction waves

To see that rarefaction curves lie along integral curves, and to explicitly determine the function $w(x/t)$ in (8.1), we differentiate $u(x, t) = w(x/t)$ to obtain

$$u_t(x, t) = -\frac{x}{t^2} w'(x/t) \tag{8.3}$$

$$u_x(x,t) \;=\; \frac{1}{t}w'(x/t) \tag{8.4}$$

so that $u_t + f'(u)u_x = 0$ yields

$$-\frac{x}{t^2}w'(x/t) + \frac{1}{t}f'(w(x/t))w'(x/t) = 0. \tag{8.5}$$

Multiplying by t and rearranging gives

$$f'(w(\xi))w'(\xi) = \xi w'(\xi), \tag{8.6}$$

where $\xi = x/t$. one possible solution of (8.6) is $w'(\xi) \equiv 0$, *i.e.*, w constant. Any constant function is a similarity solution of the conservation law, and indeed the rarefaction wave (8.1) takes this form for $\xi < \xi_1$ and $\xi > \xi_2$. In between, w is presumably smoothly varying and $w' \neq 0$. Then (8.6) says that $w'(\xi)$ must be proportional to some eigenvector $r_p(w(\xi))$ of $f'(w(\xi))$,

$$w'(\xi) = \alpha(\xi)r_p(w(\xi)) \tag{8.7}$$

and hence the values $w(\xi)$ all lie along some integral curve of r_p. In particular, the states $u_l = w(\xi_1)$ and $u_r = w(\xi_2)$ both lie on the same integral curve. This is a necessary condition for the existence of a rarefaction wave connecting u_l and u_r, but note that it is not sufficient. We need $\xi = x/t$ to be monotonically increasing as $w(\xi)$ moves from u_l to u_r along the integral curve; otherwise the function (8.1) is not single-valued. Note that our parameterization of the integral curve by ξ is not at all arbitrary at this point, since (8.6) requires that ξ be an eigenvalue of $f'(w(\xi))$,

$$\xi = \lambda_p(w(\xi)). \tag{8.8}$$

This particular parameterization is forced by our definition $\xi = x/t$. Note that (8.8) implies that w is constant along the ray $x = \lambda_p(w)t$, and hence each constant value of w propagates with speed $\lambda_p(w)$, just as in the scalar problem.

By (8.8), monotonicity of ξ is equivalent to monotonicity of $\lambda_p(w)$ as w moves from u_l to u_r. From a given state u_l we can move along the integral curve only in the direction in which λ_p is increasing. If λ_p has a local maximum at u_l in the direction r_p, then there are no rarefaction waves with left state u_l. In the generic nonlinear case, there is a one parameter family of states that can be connected to u_l by a p-rarefaction – all those states lying on the integral curve of r_p in the direction of increasing λ_p up to the next local maximum of λ_p.

If the pth field is genuinely nonlinear then λ_p is monotonically varying along the entire integral curve. We need not worry about local maxima and we see that u_l and u_r can always be connected by a rarefaction wave provided they lie on the same integral curve and

$$\lambda_p(u_l) < \lambda_p(u_r). \tag{8.9}$$

If the pth field is linearly degenerate, then λ_p is constant on each integral curve and there are no possible rarefaction waves in this family.

In order to explicitly determine the function $w(\xi)$, we first determine the scale factor $\alpha(\xi)$ in (8.7) by differentiating (8.8) with respect to ξ. This gives

$$\begin{aligned} 1 &= \nabla\lambda_p(w(\xi)) \cdot w'(\xi) \\ &= \alpha(\xi)\,\nabla\lambda_p(w(\xi)) \cdot r_p(w(\xi)) \end{aligned}$$

using (8.7), and hence

$$\alpha(\xi) = \frac{1}{\nabla\lambda_p(w(\xi)) \cdot r_p(w(\xi))}. \tag{8.10}$$

Using this in (8.7) gives a system of ordinary differential equations for $w(\xi)$:

$$w'(\xi) = \frac{r_p(w(\xi))}{\nabla\lambda_p(w(\xi)) \cdot r_p(w(\xi))}, \qquad \xi_1 \le \xi \le \xi_2 \tag{8.11}$$

with initial data

$$w(\xi_1) = u_l$$

where $\xi_1 = \lambda_p(u_l)$ and $\xi_2 = \lambda_p(u_r)$. Note that the denominator in (8.11) is finite for $\xi_1 \le \xi \le \xi_2$ only if λ_p is monotone between ξ_1 and ξ_2.

EXAMPLE 8.2. We will construct 1-rarefactions for the isothermal equations. Using (7.7) and (7.19), the system of ODEs (8.11) takes the form

$$\begin{aligned} \rho'(\xi) &= -\rho(\xi)/a, & \rho(\xi_1) &= \rho_l \\ m'(\xi) &= \rho(\xi) - m(\xi)/a, & m(\xi_1) &= m_l \end{aligned} \tag{8.12}$$

where $\xi_1 = \lambda_1(u_l) = m_l/\rho_l - a$. The first ODE is decoupled from the second and has solution

$$\rho(\xi) = \rho_l e^{-(\xi-\xi_1)/a}. \tag{8.13}$$

The second ODE is then

$$m'(\xi) = \rho_l e^{-(\xi-\xi_1)/a} - m(\xi)/a, \qquad m(\xi_1) = m_l \tag{8.14}$$

with solution

$$\begin{aligned} m(\xi) &= (\rho_l(\xi - \xi_1) + m_l)\, e^{-(\xi-\xi_1)/a} \\ &= \rho_l(\xi + a)e^{-(\xi-\xi_1)/a}. \end{aligned} \tag{8.15}$$

From the solutions $(\rho(\xi), m(\xi))$ it is also useful to eliminate ξ and solve for m as a function of ρ. This gives explicit expressions for the integral curves in the phase plane. If we solve for ξ in (8.13) and use this in (8.15) we obtain

$$m(\rho) = \rho m_l/\rho_l - a\rho\log(\rho/\rho_l). \tag{8.16}$$

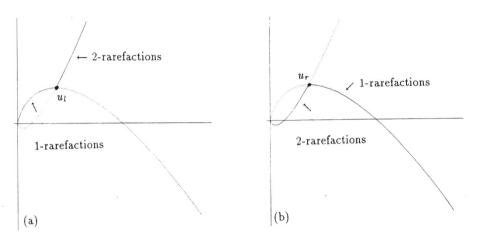

Figure 8.2. a) Set of states that can be connected to $u_l = \hat{u}$ by a rarefaction wave. b) Set of states that can be connected to $u_r = \hat{u}$ by a rarefaction wave. In each case the full integral curves are shown as dashed lines.

We can construct 2-rarefactions in exactly the same manner, obtaining

$$\rho(\xi) = \rho_l e^{(\xi - \xi_1)/a}, \qquad (8.17)$$
$$m(\xi) = \rho_l(\xi - a)e^{(\xi - \xi_1)/a}. \qquad (8.18)$$

and consequently

$$m(\rho) = \rho m_l/\rho_l + a\rho \log(\rho/\rho_l). \qquad (8.19)$$

EXERCISE 8.1. *Derive the expressions for 2-rarefactions displayed above.*

EXERCISE 8.2. *From (8.16) and (8.19) verify that $m'(\rho_l) = \lambda_p(u_l)$ and explain why this should be so.*

For a given state $\hat{u} = u_l$ we can plot the set of all states u_r which can be connected to u_l by a rarefaction wave in either the first or second family. This is shown in Figure 8.2a for a particular choice of u_l. Note that if we consider this same state $\hat{u}$ to be u_r and now plot the set of all states u_l that can be connected to $\hat{u} = u_r$ by a rarefaction, we obtain a different picture as in Figure 8.2b. We must now have ξ *decreasing* as we move away from u_r and so it is the opposite side of each integral curve that is now relevant.

Note that these integral curves are very similar to the Hugoniot locus, e.g., Figure 7.3. Locally, near the point $\hat{u}$, they must in fact be very similar. We know already that in the pth family each of these curves is tangent to $r_p(\hat{u})$ at $\hat{u}$. Moreover, it can be shown that the curvature of both curves is the same (See Lax[45]).

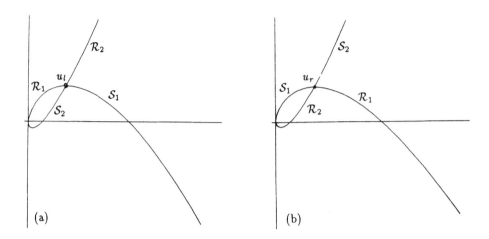

Figure 8.3. *a) Set of states that can be connected to u_l by an entropy-satisfying 1-wave or 2-wave. b) Set of states that can be connected to u_r. In each case, $\mathcal{R}_p$ denotes p-rarefactions and $\mathcal{S}_p$ denotes p-shocks.*

EXERCISE 8.3. *Verify this for the isothermal equations by computing $m''(\hat{\rho})$ for the Hugoniot locus from (7.9) and for the integral curves from (8.16) or (8.19) and seeing that they are the same. Also verify that the third derivatives are not equal.*

8.3   General solution of the Riemann problem

We can combine Figures 7.3 and 8.2 to obtain a plot showing all states that can be connected to a given $\hat{u}$ by entropy-satisfying waves, either shocks or rarefactions. Again, the nature of this plot depends on whether $\hat{u}$ is to be the left state or right state, so we obtain two plots as shown in Figure 8.3. Here $\mathcal{S}_p$ is used to denote the locus of states that can be connected by a p-shock and $\mathcal{R}_p$ is the locus of states that can be connected by a p-rarefaction. Notice that the shock and rarefaction curves match up smoothly (with the same slope and curvature) at the point $\hat{u}$.

To solve the general Riemann problem between two different states u_l and u_r, we simply superimpose the appropriate plots and look for the intersection u_m of a 1-wave curve from u_l and a 2-wave curve from u_r. An example for the isothermal equations is shown in Figure 8.4. This is the same example considered in Figure 7.2. We now see that the entropy-satisfying weak solution consists of a 1-shock from u_l to u_m followed by a 2-rarefaction from u_m to u_r.

To analytically determine the state u_m, we must first determine whether each wave is a shock or rarefaction, and then use the appropriate expressions relating m and ρ along

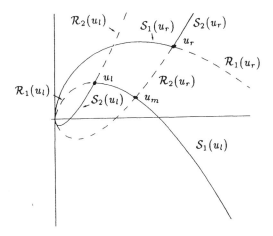

Figure 8.4. Construction of the entropy-satisfying weak solution to the Riemann problem
with left and right states u_l and u_r.

each curve to solve for the intersection. We have already seen how to do this for the case
of two shocks, by solving the equation (7.14). If the solution consists of two rarefactions
then the intermediate state must satisfy

$$m_m = \rho_m m_l/\rho_l - a\rho_m \log(\rho_m/\rho_l) \qquad (8.20)$$
$$m_m = \rho_m m_r/\rho_r + a\rho_m \log(\rho_m/\rho_r). \qquad (8.21)$$

Equating the two right hand sides gives a single equation for ρ_m alone, with solution

$$\rho_m = \sqrt{\rho_l \rho_r} \, \exp\left(\frac{1}{2a}\left(\frac{m_l}{\rho_l} - \frac{m_r}{\rho_r}\right)\right). \qquad (8.22)$$

We then obtain m_m from either (8.20) or (8.21).

If the solution consists of one shock and one rarefaction wave, as in Figure 8.4, then
we must solve for ρ_m and m_m from the equations

$$m_m = \frac{\rho_m m_l}{\rho_l} - a\sqrt{\frac{\rho_m}{\rho_l}}\,(\rho_m - \rho_l) \qquad (8.23)$$
$$m_m = \frac{\rho_m m_r}{\rho_r} + a\rho_m \log(\rho_m/\rho_l),$$

for example, in the case of a 1-shock followed by a 2-rarefaction. In this case it is not
possible to obtain a closed form solution (ρ_m, m_m). Instead, it is necessary to solve these
two equations by an iterative method such as Newton's method.

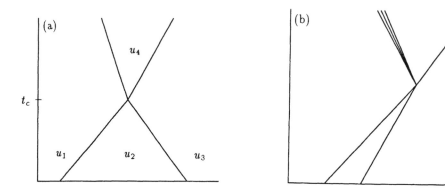

Figure 8.5. (a) Collision of a 2-shock and a 1-shock. (b) Collision of two 2-shocks.

8.4   Shock collisions

When two shocks collide the two discontinuities merge into one discontinuity that no longer satisfies the R-H condition. We can determine what happens beyond this point by solving a new Riemann problem. Consider Figure 8.5a where we have initial data consisting of three constant states u_1, u_2 and u_3 chosen such that u_1 and u_2 are connected by a 2-shock while u_2 and u_3 are connected by a 1-shock. At the collision time t_c, the function $u(x, t_c)$ has only a single discontinuity from u_1 to u_3. The behavior beyond this point is determined by solving the Riemann problem with $u_l = u_1$ and $u_r = u_3$.

EXERCISE 8.4. *Consider the isothermal equations with $a = 1$ and take data*

$$u(x,0) = \begin{cases} u_1 & x < 0.5 \\ u_2 & 0.5 < x < 1 \\ u_3 & x > 1 \end{cases}$$

where

$$u_1 = \begin{bmatrix} .28260 \\ .098185 \end{bmatrix} \qquad u_2 = \begin{bmatrix} .2 \\ 0 \end{bmatrix} \qquad u_3 = \begin{bmatrix} .35 \\ -0.19843 \end{bmatrix}.$$

Determine the solution $u(x, t)$.

EXERCISE 8.5. *For the isothermal equations, use the structure of the integral curves and Hugoniot loci to argue that:*

1. *When two shocks of different families collide, the result is again two shocks.*

2. *When two shocks in the same family collide, the result is a shock in that family and a rarefaction wave in the other family. (see Figure 8.5b).*

9 The Riemann problem for the Euler equations

I will not attempt to present all of the details for the case of the Euler equations. In principle we proceed as in the examples already presented, but the details are messier. Instead, I will concentrate on discussing one new feature seen here, contact discontinuities, and see how we can take advantage of the linear degeneracy of one field to simplify the solution process for a general Riemann problem. Full details are available in many sources, for example [11], [77], [97].

If we compute the Jacobian matrix $f'(u)$ from (2.14), with the polytropic equation of state (5.22), we obtain

$$f'(u) = \begin{bmatrix} 0 & 1 & 0 \\ \frac{1}{2}(\gamma - 3)v^2 & (3 - \gamma)v & (\gamma - 1) \\ \frac{1}{2}(\gamma - 1)v^3 - v(E + p)/\rho & (E + p)/\rho - (\gamma - 1)v^2 & \gamma v \end{bmatrix}. \qquad (9.1)$$

The eigenvalues are

$$\lambda_1(u) = v - c, \quad \lambda_2(u) = v, \quad \lambda_3(u) = v + c \qquad (9.2)$$

where c is the sound speed,

$$c = \sqrt{\frac{\gamma p}{\rho}}. \qquad (9.3)$$

9.1 Contact discontinuities

Of particular note in these equations is the fact that the second characteristic field is linearly degenerate. It is easy to check from (9.1) that

$$r_2(u) = \begin{bmatrix} 1 \\ v \\ \frac{1}{2}v^2 \end{bmatrix} \qquad (9.4)$$

is an eigenvector of $f'(u)$ with eigenvalue $\lambda_2(u) = v = (\rho v)/\rho$. Since

$$\nabla \lambda_2(u) = \begin{bmatrix} -v/\rho \\ 1/\rho \\ 0 \end{bmatrix} \tag{9.5}$$

we find that $\nabla \lambda_2 \cdot r_2 \equiv 0$.

The first and third fields are genuinely nonlinear (somewhat more tedious to check).

Since the second field is linearly degenerate, we can have neither rarefaction waves nor shocks in this field. Instead we have contact discontinuities, which are linear discontinuities that propagate with speed equal to the characteristic speed λ_2 on each side.

Note that because $\lambda_2 = v$ is constant on the integral curves of r_2, and since r_2 depends only on v, the vector r_2 is itself constant on these curves, and hence the integral curves are straight lines in phase space. Moreover, these integral curves also form the Hugoniot locus for contact discontinuities. To see this, consider the Rankine-Hugoniot condition between states u and $\hat{u}$:

$$\begin{aligned} \rho v - \hat{\rho} \hat{v} &= s(\rho - \hat{\rho}) \\ \left(\frac{1}{2}\rho v^2 + p\right) - \left(\frac{1}{2}\hat{\rho}\hat{v}^2 + \hat{p}\right) &= s(\rho v - \hat{\rho}\hat{v}) \\ v(E + p) - \hat{v}(\hat{E} + \hat{p}) &= s(E - \hat{E}). \end{aligned} \tag{9.6}$$

These equations are clearly satisfied if we take $s = v = \hat{v}$ and $p = \hat{p}$. But then

$$u - \hat{u} = \begin{bmatrix} \rho - \hat{\rho} \\ \rho v - \hat{\rho}\hat{v} \\ \left(\frac{1}{2}\rho v^2 + p/(\gamma - 1)\right) - \left(\frac{1}{2}\hat{\rho}\hat{v}^2 + p/(\gamma - 1)\right) \end{bmatrix} = (\rho - \hat{\rho})r_2(u). \tag{9.7}$$

The jump is in the direction $r_2(u)$ and so the Hugoniot locus coincides with the integral curve of r_2.

We see from this that across a contact discontinuity the velocity v and pressure p are constant, only the density jumps (with resulting jumps in the other conserved quantities momentum and energy). Note that this is a special case of (6.32). Also notice that the speed of propagation is the same as the fluid velocity v. Thus particle paths agree with the characteristics, and move parallel to the contact discontinuity.

Since particle paths do not cross the discontinuity, two gases which are initially in contact with one another and have the same velocity and pressure will stay distinct, and continue to be in contact along this discontinuity.

It may seem strange that this discontinuity can sustain a jump in density — it seems that the denser gas should try to expand into the thinner gas. But that's because our intuition tends to equate higher density with higher pressure. It is only a *pressure* difference that can provide the force for expansion, and here the pressures are equal.

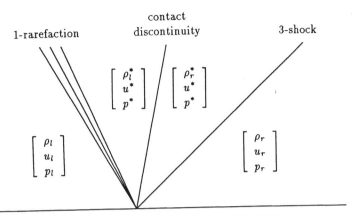

Figure 9.1. Typical solution to the Riemann problem for the Euler equations.

We can achieve two different densities at the same pressure by taking gases at two different temperatures. In fact, from (5.21) it is clear that there must be a jump in temperature if $\rho \neq \hat{\rho}$ while $p = \hat{p}$. There must also be a jump in entropy by (5.23). This explains why contact discontinuities do not appear in solutions to the isothermal or isentropic equations considered previously. In the reduction of the Euler equations to one of these systems of only two equations, it is this linearly degenerate characteristic field that disappears.

9.2 Solution to the Riemann problem

The first and third characteristic fields are genuinely nonlinear and have behavior similar to the two characteristic fields in the isothermal equations. The solution to a Riemann problem typically has a contact discontinuity and two nonlinear waves, each of which might be either a shock or a rarefaction wave depending on u_l and u_r. A typical solution is shown in Figure 9.1.

Because v and p are constant across the contact discontinuity, it is often easier to work in the variables (ρ, v, p) rather than $(\rho, \rho v, E)$, although of course the jump conditions must be determined using the conserved variables. The resulting Hugoniot locus and integral curves can be transformed into (ρ, v, p) space.

If the Riemann data is (ρ_l, v_l, p_l) and (ρ_r, v_r, p_r), then the two new constant states that appear in the Riemann solution will be denoted by $u_l^* = (\rho_l^*, v^*, p^*)$ and $u_r^* = (\rho_r^*, v^*, p^*)$. (See Figure 9.1.) Note that across the 2-wave we know there is a jump only in density.

Solution of the Riemann problem proceeds in principle just as in the previous chapter. Given the states u_l and u_r in the phase space, we need to determine the two intermediate states in such a way that u_l and u_l^* are connected by a 1-wave, u_l^* and u_r^* are connected

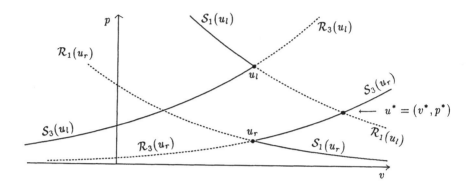

Figure 9.2.　Projection of shock and rarefaction curves onto the two-dimensional v–p plane, and determination of u^*.

by a 2-wave, and finally u_r^* and u_r are connected by a 3-wave.

　　This seems difficult, but we can take advantage of the fact that we know the 2-wave will be a contact discontinuity across which v and p are constant to make the problem much simpler. Instead of considering the full three dimensional (ρ, v, p) phase space, consider the v–p plane and project the integral curves and Hugoniot loci for the 1-waves and 3-waves onto this plane. In particular, project the locus of all states that can be connected to u_l by a 1-wave (entropy satisfying shocks or rarefactions) onto this plane and also the locus of all states that can be connected to u_r by a 3-wave. This gives Figure 9.2.

　　We see in this example that we can go from u_l (or actually, the projection of u_l) to u^* by a 1-rarefaction and from u^* to u_r by a 3-shock. The problem with this construction, of course, is that these curves are really curves in 3-space, and just because their projections intersect does not mean the original curves intersect. However, the curve $\mathcal{R}_1(u_l)$ must go through some state $u_l^* = (\rho_l^*, v^*, p^*)$ for some ρ_l^* (so that it's projection onto the v–p plane is (v^*, p^*)). Similarly, the curve $\mathcal{S}_3(u_r)$ must pass through some state $u_r^* = (\rho_r^*, v^*, p^*)$. But these two states differ only in ρ, and hence can be connected by a 2-wave (contact discontinuity). We have thus achieved our objective. Note that this technique depends on the fact that *any* jump in ρ is allowed across the contact discontinuity.

　　In practice the calculation of u^* can be reduced to a single nonlinear equation for p^*, which is solved by an iterative method. Once p^* is known, u^*, ρ_l^* and ρ_r^* are easily determined. Godunov first proposed a numerical method based on the solution of Riemann problems and presented one such iterative method in his paper[24] (also described in §12.15 of [63]). Chorin[6] describes an improvement of this method. More details on the solution of the Riemann problem can also be found in §81 of [11].

Part II

Numerical Methods

10 Numerical Methods for Linear Equations

Before studying numerical methods for nonlinear conservation laws, we review some of the basic theory of numerical methods for the linear advection equation and linear hyperbolic systems. The emphasis will be on concepts that carry over to the nonlinear case.

We consider the time-dependent Cauchy problem in one space dimension,

$$u_t + Au_x = 0, \qquad -\infty < x < \infty, \quad t \geq 0, \tag{10.1}$$
$$u(x,0) = u_0(x).$$
$$\tag{10.2}$$

We discretize the x-t plane by choosing a **mesh width** $h \equiv \Delta x$ and a **time step** $k \equiv \Delta t$, and define the discrete mesh points (x_j, t_n) by

$$x_j = jh, \quad j = \ldots, -1, 0, 1, 2, \ldots$$
$$t_n = nk, \quad n = 0, 1, 2, \ldots$$

It will also be useful to define

$$x_{j+1/2} = x_j + h/2 = \left(j + \frac{1}{2}\right)h.$$

For simplicity we take a uniform mesh, with h and k constant, although most of the methods discussed can be extended to variable meshes.

The finite difference methods we will develop produce approximations $U_j^n \in \mathbb{R}^m$ to the solution $u(x_j, t_n)$ at the discrete grid points. The pointwise values of the true solution will be denoted by

$$u_j^n = u(x_j, t_n).$$

This is a standard interpretation of the approximate solution, and will be used at times here, but in developing methods for conservation laws it is often preferable to view U_j^n as an approximation to a **cell average** of $u(x, t_n)$, defined by

$$\bar{u}_j^n \equiv \frac{1}{h} \int_{x_{j-1/2}}^{x_{j+1/2}} u(x, t_n)\, dx \tag{10.3}$$

rather than as an approximation to the pointwise value u_j^n. This interpretation is natural since the integral form of the conservation law describes precisely the time evolution of integrals such as that appearing in (10.3).

As initial data for the numerical method we use $u_0(x)$ to define U^0 either by pointwise values, $U_j^0 = u_j^0$, or preferably by cell averages, $U_j^0 = \bar{u}_j^0$.

It is also frequently convenient to define a piecewise constant function $U_k(x, t)$ for all x and t from the discrete values U_j^n. We assign this function the value U_j^n in the (j, n) grid cell, $i.e.$,

$$U_k(x, t) = U_j^n \quad \text{for } (x, t) \in [x_{j-1/2}, x_{j+1/2}) \times [t_n, t_{n+1}). \tag{10.4}$$

We index this function U_k by the time step k, and assume that the mesh width h and time step k are related in some fixed way, so that the choice of k defines a unique mesh. For time-dependent hyperbolic equations one generally assumes that the **mesh ratio** k/h is a fixed constant as k, $h \to 0$. This assumption will be made from here on.

We will primarily study the theory of numerical methods for the Cauchy problem, as indicated in (10.1). In practice we must compute on a finite spatial domain, say $a \leq x \leq b$, and we require appropriate boundary conditions at a and/or b. One simple case is obtained if we take **periodic boundary conditions,** where

$$u(a, t) = u(b, t) \qquad \forall\, t \geq 0 \tag{10.5}$$

is specified. Then we can use periodicity in applying finite difference methods as well. This is equivalent to a Cauchy problem with periodic initial conditions, since the solution remains periodic and we need compute over only one period.

For linear equations, the study of the Cauchy problem or periodic case is particularly attractive since Fourier transform techniques can be used. The study of the general initial boundary value problem is more complex.

From the initial data $u_0(x)$ we have defined data U^0 for our approximate solution. We now use a **time-marching** procedure to construct the approximation U^1 from U^0, then U^2 from U^1 (and possibly also U^0) and so on. In general, we construct U^{n+1} from U^n in a two-level method, or more generally from U^n, U^{n-1}, $\ldots$, U^{n-r} in an $(r + 2)$-level method.

There are a wide variety of finite difference methods that can be used. Many of these are derived simply by replacing the derivatives occurring in (10.1) by appropriate finite difference approximations. For example, replacing u_t by a forward-in-time approximation and u_x by a spatially centered approximation, we obtain the following difference equations for U^{n+1}:

$$\frac{U_j^{n+1} - U_j^n}{k} + A\left(\frac{U_{j+1}^n - U_{j-1}^n}{2h}\right) = 0. \qquad (10.6)$$

This can be solved for U_j^{n+1} to obtain

$$U_j^{n+1} = U_j^n - \frac{k}{2h}A(U_{j+1}^n - U_{j-1}^n). \qquad (10.7)$$

Unfortunately, despite the quite natural derivation of this method, it suffers from severe stability problems and is useless in practice.

A far more stable method is obtained by evaluating the centered difference approximations to u_x at time t_{n+1} rather than at time t_n, giving

$$\frac{U_j^{n+1} - U_j^n}{k} + A\left(\frac{U_{j+1}^{n+1} - U_{j-1}^{n+1}}{2h}\right) = 0 \qquad (10.8)$$

or

$$U_j^{n+1} = U_j^n - \frac{k}{2h}A(U_{j+1}^{n+1} - U_{j-1}^{n+1}). \qquad (10.9)$$

Note, however, that in order to determine U^{n+1} from the data U^n in each time step, we must view equation (10.9) as a coupled system of equations over all values of j. This would be an infinite system for the Cauchy problem, but in any practical calculation we would use a bounded interval with N grid points and this would be a finite system. In the scalar case this is an $N \times N$ linear system with a tridiagonal coefficient matrix. When $u \in \mathbb{R}^m$, the discrete system is $mN \times mN$.

The method (10.7) allows us to determine U^{n+1} explicitly, and is called an **explicit** method, whereas (10.9) is an **implicit** method. Implicit methods are rarely used for time-dependent hyperbolic problems, although they are very important for other classes of equations. Although the method (10.7) is useless due to stability problems, there are other explicit methods which work very satisfactorily with reasonable size time steps and these are typically more efficient than implicit methods.

If we look at which grid points are involved in the computation of U_j^{n+1} with a given method, we can obtain a diagram that is known as the **stencil** of the method. The stencils for the methods (10.7) and (10.9) are shown in Figure 10.1.

A wide variety of methods can be devised for the linear system (10.1) by using different finite difference approximations. A few possibilities are listed in Table 10.1, along with their stencils. Most of these are based directly on finite difference approximations to the PDE. An exception is the Lax-Wendroff method, which is based on the Taylor series expansion

$$u(x, t + k) = u(x, t) + ku_t(x, t) + \frac{1}{2}k^2 u_{tt}(x, t) + \cdots \qquad (10.10)$$

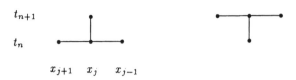

t_{n+1}

t_n

$x_{j+1} \quad x_j \quad x_{j-1}$

Figure 10.1. Stencils for the methods (10.7) and (10.9).

and the observation that from $u_t = -Au_x$ we can compute

$$u_{tt} = -Au_{xt} = -Au_{tx} = -A(-Au_x)_x = A^2 u_{xx} \tag{10.11}$$

so that (10.10) becomes

$$u(x, t + k) = u(x, t) - kAu_x(x, t) + \frac{1}{2}k^2 A^2 u_{xx}(x, t) + \cdots \tag{10.12}$$

The Lax-Wendroff method then results from retaining only the first three terms of (10.12) and using centered difference approximations for the derivatives appearing there.

The Beam-Warming method is a one-sided version of Lax-Wendroff. It is also obtained from (10.12), but now using second order accurate one-sided approximations of the derivatives.

With the exception of the Leapfrog method, all of the methods in Table 10.1 are 2-level methods. For time-dependent conservation laws, 2-level methods are almost exclusively used. Methods involving more than 2 levels have additional difficulties. They require more storage (a disadvantage primarily in two or three dimensional calculations) and they require special startup procedures, since initially only one level of data is known. We will study explicit 2-level methods almost exclusively, and introduce some special notation for such methods, writing

$$U^{n+1} = \mathcal{H}_k(U^n). \tag{10.13}$$

Here U^{n+1} represents the vector of approximations U_j^{n+1} at time t_{n+1}. The value U_j^{n+1} at a particular point j typically depends on several values from the vector U^n (depending on how large the stencil is), and so we write

$$U_j^{n+1} = \mathcal{H}_k(U^n; j) \tag{10.14}$$

to signify that U_j^{n+1} depends on the full vector U^n. For example, the $\mathcal{H}_k$ operator for method (10.7) takes the form

$$\mathcal{H}_k(U^n; j) = U_j^n - \frac{k}{2h}A(U_{j+1}^n - U_{j-1}^n). \tag{10.15}$$

Name	Difference Equations	Stencil
Backward Euler	$U_j^{n+1} = U_j^n - \frac{k}{2h}A(U_{j+1}^n - U_{j-1}^n)$	
One-sided	$U_j^{n+1} = U_j^n - \frac{k}{h}A(U_j^n - U_{j-1}^n)$	
One-sided	$U_j^{n+1} = U_j^n - \frac{k}{h}A(U_{j+1}^n - U_j^n)$	
Lax-Friedrichs	$U_j^{n+1} = \frac{1}{2}(U_{j-1}^n + U_{j+1}^n) - \frac{k}{2h}A(U_{j+1}^n - U_{j-1}^n)$	
Leapfrog	$U_j^{n+1} = U_j^{n-1} - \frac{k}{2h}A(U_{j+1}^n - U_{j-1}^n)$	
Lax-Wendroff	$U_j^{n+1} = U_j^n - \frac{k}{2h}A(U_{j+1}^n - U_{j-1}^n)$ $+ \frac{k^2}{2h^2}A^2(U_{j+1}^n - 2U_j^n + U_{j-1}^n)$	
Beam-Warming	$U_j^{n+1} = U_j^n - \frac{k}{2h}A(3U_j^n - 4U_{j-1}^n + U_{j-2}^n)$ $+ \frac{k^2}{2h^2}A^2(U_j^n - 2U_{j-1}^n + U_{j-2}^n)$	

TABLE 10.1

Finite difference methods for the linear problem $u_t + Au_x = 0$.

The finite difference operator $\mathcal{H}_k$ can be extended to apply to functions of x rather than to discrete grid functions in a very natural way. If $v(x)$ is an arbitrary function of x then we define $\mathcal{H}_k(v)$ to be the new function of x obtained by applying the difference scheme to v at each point x. Pointwise values of $\mathcal{H}_k(v)$ might be denoted by $[\mathcal{H}_k(v)](x)$, or, for consistency with (10.14), by $\mathcal{H}_k(v; x)$. For example, applying the method (10.7) to $v(x)$ gives $\mathcal{H}_k(v)$ defined by

$$\mathcal{H}_k(v; x) = [\mathcal{H}_k(v)](x) = v(x) - \frac{k}{2h} A(v(x+h) - v(x-h)). \tag{10.16}$$

Notice in particular that if we apply this operator to the piecewise constant function $U_k(\cdot, t)$ (defined by our discrete values via (10.4)) at any time t, we obtain the piecewise constant function $U_k(\cdot, t+k)$, so[1]

$$U_k(x, t+k) = \mathcal{H}_k(U_k(\cdot, t); x). \tag{10.17}$$

This demonstrates a certain consistency in our notation, and we will use the same symbol $\mathcal{H}_k$ to denote both the discrete and continuous operators.

Notice that all of the methods listed in Table 10.1 are **linear methods**, meaning that $\mathcal{H}_k$ is a linear operator,

$$\mathcal{H}_k(\alpha U^n + \beta V^n) = \alpha \mathcal{H}_k(U^n) + \beta \mathcal{H}_k(V^n) \tag{10.18}$$

for any grid functions U^n and V^n and scalar constants α, β. Linearity of difference methods is heavily used in the study of discrete approximations to linear PDEs. If $\mathcal{H}_k$ is linear then we can write (10.13) in the form

$$U^{n+1} = \mathcal{H}_k U^n \tag{10.19}$$

where $\mathcal{H}_k$ is now an $mN \times mN$ matrix in the discrete case (when solving a system of m equations on a grid with N grid points).

10.1   The global error and convergence

We are ultimately interested in how well U_j^n approximates the true solution, and we define the **global error** to be the difference between the true and computed solutions. We can make this more precise in several ways. In studying smooth solutions it is usually most convenient to consider the pointwise error,

$$E_j^n = U_j^n - u_j^n. \tag{10.20}$$

---

[1]A note on notation: $U_k(\cdot, t)$ denotes the function of x alone obtained by fixing t. The operator $\mathcal{H}_k$ applies only to functions of one variable.

For conservation laws it is sometimes preferable to consider the error relative to the cell average of the true solution,

$$\bar{E}_j^n = U_j^n - \bar{u}_j^n. \tag{10.21}$$

These can be unified to some extent by introducing the error function

$$E_k(x, t) = U_k(x, t) - u(x, t). \tag{10.22}$$

Then E_j^n is the pointwise value $E_k(x_j, t_n)$ while $\bar{E}_j^n$ is the cell average of E_k at time t_n.

With these definitions, we say that a method is **convergent** in some particular norm $\|\cdot\|$ if

$$\|E_k(\cdot, t)\| \to 0 \quad \text{as } k \to 0, \tag{10.23}$$

for any fixed $t \geq 0$, and for all initial data u_0 in some class.

10.2 Norms

In discussing convergence we must make a choice of norm. It may happen that a method is convergent in one norm but not in another. For conservation laws the natural norm to use is the 1-norm, defined for a general function $v(x)$ by

$$\|v\|_1 = \int_{-\infty}^{\infty} |v(x)| \, dx, \tag{10.24}$$

so that

$$\|E_k(\cdot, t)\|_1 = \int_{-\infty}^{\infty} |E_k(x, t)| \, dx$$

at each fixed t. This norm is natural since it requires essentially just integrating the function, and the form of the conservaion law often allows us to say something about these integrals.

Ideally one might hope for convergence in the ∞-norm,

$$\|v\|_\infty = \sup_x |v(x)|, \tag{10.25}$$

but this is unrealistic in approximating discontinuous solutions. The pointwise error in the neighborhood of a discontinuity typically does not go to zero uniformly as the grid is refined even though the numerical results may be perfectly satisfactory. The method may well converge in the 1-norm but not in the ∞-norm.

For linear equations a more popular norm is the 2-norm,

$$\|v\|_2 = \left[\int_{-\infty}^{\infty} |v(x)|^2 \, dx \right]^{1/2}, \tag{10.26}$$

since for linear problems Fourier analysis is applicable and Parseval's relation says that the Fourier transform $\hat{v}(\xi)$ of $v(x)$ has the same 2-norm as v. This can be used to simplify the analysis of linear methods considerably.

We will use the 1-norm almost exclusively, and so a norm with no subscript will generally refer to the 1-norm. For the discrete grid function U^n we use the discrete 1-norm defined by

$$\|U^n\|_1 = h \sum_j |U_j^n|. \tag{10.27}$$

Note that this is consistent with the function version in the sense that

$$\|U^n\|_1 = \|U_k(\cdot, t_n)\|_1.$$

Although our ultimate interest is in proving convergence, and better yet obtaining explicit bounds on the size of the global error (and hence convergence *rates*), this is hard to achieve directly. Instead we begin by examining the *local truncation error* and then use *stability* of the method to derive estimates for the global error from the local error.

10.3   Local truncation error

The local truncation error $L_k(x, t)$ is a measure of how well the difference equation models the differential equation locally. It is defined by replacing the approximate solution U_j^n in the difference equations by the true solution $u(x_j, t_n)$. Of course this true solution of the PDE is only an approximate solution of the difference equations, and how well it satisfies the difference equations gives an indication of how well the exact solution of the difference equations satisfies the differential equation.

As an example, consider the **Lax-Friedrichs** method from Table 10.1. This method is similar to the unstable method (10.7) but replaces U_j^n by $\frac{1}{2}(U_{j-1}^n + U_{j+1}^n)$ and is stable provided k/h is sufficiently small, as we will see later.

We first write this method in the form

$$\frac{1}{k}\left[U_j^{n+1} - \frac{1}{2}(U_{j-1}^n + U_{j+1}^n)\right] + \frac{1}{2h}A[U_{j+1}^n - U_{j-1}^n] = 0,$$

so that it appears to be a direct discretization of the PDE. If we now replace each U_j^n by the exact solution at the corresponding point, we will not get zero exactly. What we get instead is defined to be the local truncation error,

$$
\begin{aligned}
L_k(x, t) &= \frac{1}{k}\left[u(x, t+k) - \frac{1}{2}(u(x-h, t) + u(x+h, t))\right] \\
&\quad + \frac{1}{2h}A[u(x+h, t) - u(x-h, t)].
\end{aligned} \tag{10.28}
$$

In computing the local truncation error we always assume *smooth* solutions, and so we can expand each term on the right hand side of (10.28) in a Taylor series about $u(x, t)$.

Doing this and collecting terms gives (with $u \equiv u(x,t)$):

$$L_k(x,t) = \frac{1}{k}\left[(u + ku_t + \frac{1}{2}k^2 u_{tt} + \cdots) - (u + \frac{1}{2}h^2 u_{xx} + \cdots)\right]$$
$$+ \frac{1}{2h}A\left[2hu_x + \frac{1}{3}h^3 u_{xxx} + \cdots\right]$$
$$= u_t + Au_x + \frac{1}{2}\left(ku_{tt} - \frac{h^2}{k}u_{xx}\right) + O(h^2). \tag{10.29}$$

Since we assume that $u(x,t)$ is the exact solution, $u_t + Au_x = 0$ in (10.29). Using this and also (10.11), we find that

$$L_k(x,t) = \frac{1}{2}k\left(A^2 - \frac{h^2}{k^2}I\right)u_{xx}(x,t) + O(k^2)$$
$$= O(k) \quad \text{as } k \to 0. \tag{10.30}$$

Recall that we assume a fixed relation between k and h, $k/h = $ constant, so that h^2/k^2 is constant as the mesh is refined. (This also justifies indexing L_k by k alone rather than by both k and h.)

By being more careful in this analysis, using Taylor's theorem with remainder and assuming uniform bounds on the appropriate derivatives of $u(x,t)$, we can in fact show a sharp bound of the form

$$|L_k(x,t)| \le Ck \quad \text{for all } k < k_0 \tag{10.31}$$

The constant C depends only on the initial data u_0, since for the linear system we consider here we can easily bound derivatives of $u(x,t)$ in terms of derivatives of the initial data. If we assume moreover that u_0 has compact support, then $L_k(x,t)$ will have finite 1-norm at each time t and we can obtain a bound of the form

$$\|L_k(\cdot,t)\| \le C_L k \quad \text{for all } k < k_0 \tag{10.32}$$

for some constant C_L again depending on u_0.

The Lax-Friedrichs method is said to be first order accurate since the local error (10.32) depends linearly on k.

We now extend these notions to arbitrary 2-level methods.

DEFINITION 10.1. *For a general 2-level method, we define the* **local truncation error** *by*

$$L_k(x,t) = \frac{1}{k}[u(x,t+k) - \mathcal{H}_k(u(\cdot,t);x)]. \tag{10.33}$$

DEFINITION 10.2. *The method is* **consistent** *if*

$$\|L_k(\cdot,t)\| \to 0 \quad \text{as } k \to 0. \tag{10.34}$$

DEFINITION 10.3. *The method is of order p if for all sufficiently smooth initial data with compact support, there is some constant C_L such that*

$$\|L_k(\cdot, t)\| \le C_L k^p \quad \text{for all } k < k_0, \ t \le T \tag{10.35}$$

This is the *local order* of the method, but it turns out that for smooth solutions, the global error will be of the *same* order provided the method is *stable*.

10.4   Stability

Note that we can rewrite (10.33) in the form

$$u(x, t + k) = \mathcal{H}_k(u(\cdot, t); x) + k L_k(x, t). \tag{10.36}$$

Since the numerical solution satisfies (10.17), we can obtain a simple recurrence relation for the error in a *linear* method by subtracting these two expressions, obtaining

$$E_k(x, t + k) = \mathcal{H}_k(E_k(\cdot, t); x) - k L_k(x, t). \tag{10.37}$$

Note that linearity is strongly required here. Since the method is linear, we can rewrite (10.37) in the functional form

$$E_k(\cdot, t + k) = \mathcal{H}_k E_k(\cdot, t) - k L_k(\cdot, t). \tag{10.38}$$

The error (10.37) at time $t + k$ is seen to consist of two parts. One is the new local error $-k L_k$ introduced in this time step. The other part is the cumulative error from previous time steps. By applying this relation recursively we obtain an expression for the error at time t_n:

$$E_k(\cdot, t_n) = \mathcal{H}_k^n E_k(\cdot, 0) - k \sum_{i=1}^{n} \mathcal{H}_k^{n-i} L_k(\cdot, t_{i-1}). \tag{10.39}$$

Here superscripts on $\mathcal{H}_k$ represent powers of the matrix (or linear operator) obtained by repeated applications.

In order to obtain a bound on the global error, we must insure that the local error $L_k(\cdot, t_{i-1})$ is not unduly amplified by applying $n - i$ steps of the method. This requires stability of the method, and we use a form that is often referred to as **Lax-Richtmyer** stability.

DEFINITION. *The method is **stable** if for each time T there is a constant C_S and a value $k_0 > 0$ such that*

$$\|\mathcal{H}_k^n\| \le C_S \quad \text{for all } nk \le T, \ k < k_0. \tag{10.40}$$

Note in particular that the method is stable if $\|\mathcal{H}_k\| \leq 1$, for then $\|\mathcal{H}_k^n\| \leq \|\mathcal{H}_k\|^n \leq 1$ for all n, k. More generally, some growth is allowed. For example, if

$$\|\mathcal{H}_k\| \leq 1 + \alpha k \quad \text{for all } k < k_0 \tag{10.41}$$

then

$$\|\mathcal{H}_k^n\| \leq (1 + \alpha k)^n \leq e^{\alpha k n} \leq e^{\alpha T}$$

for all k, n with $nk \leq T$.

10.5 The Lax Equivalence Theorem

This is the fundamental convergence theorem for linear difference methods, and says that for a *consistent*, linear method, *stability* is necessary and sufficient for *convergence*.

A full proof may be found in [63] or [82]. Here I will just indicate why this is sufficient, at least for smooth initial data, and also show why the global error should be of the same order as the local truncation error. This follows quite easily from the expression (10.39). Taking norms and using the triangle inequality gives

$$\|E_k(\cdot, t_n)\| \leq \|\mathcal{H}_k^n\| \, \|E_k(\cdot, 0)\| + k \sum_{i-1}^{n} \|\mathcal{H}_k^{n-i}\| \, \|L_k(\cdot, t_i)\|. \tag{10.42}$$

From stability we have $\|\mathcal{H}_k^{n-i}\| \leq C_S$ for $i = 0, 1, \ldots, n$ and so

$$\|E_k(\cdot, t_n)\| \leq C_S \left(\|E_k(\cdot, 0)\| + k \sum_{i=1}^{n} \|L_k(\cdot, t_i)\| \right). \tag{10.43}$$

If the method is pth order accurate then (10.35) holds, and so we have, for $kn = t_n \leq T$,

$$\|E_k(\cdot, t_n)\| \leq C_S \left(\|E_k(\cdot, 0)\| + T C_L k^p \right). \tag{10.44}$$

If there is no error in the initial data, then this gives the desired result. More generally, we have convergence at the expected rate provided any error in the initial data also decays at the rate $O(k^p)$ as $k \to 0$. We then obtain a global error bound of the form

$$\|E_k(\cdot, t)\| \leq C_E k^p \tag{10.45}$$

for all fixed $t \leq T$ and $k < k_0$.

EXAMPLE 10.1. Consider the Lax-Friedrichs method applied to the scalar advection equation $u_t + a u_x = 0$. We will show that the method is stable provided that k and h are related in such a way that

$$\left| \frac{ak}{h} \right| \leq 1. \tag{10.46}$$

This is the **stability restriction** for the method. For the discrete operator $\mathcal{H}_k$, we will show that $\|U^{n+1}\| \leq \|U^n\|$ and hence $\|\mathcal{H}_k\| \leq 1$. Exactly the same proof carries over to obtain the same bound for the continuous operator as well.

We have

$$U_j^{n+1} = \frac{1}{2}(U_{j-1}^n + U_{j+1}^n) - \frac{ak}{2h}(U_{j+1}^n - U_{j-1}^n)$$

and hence

$$\|U^{n+1}\| = h\sum_j |U_j^{n+1}|$$

$$\leq \frac{h}{2}\left[\sum_j \left|\left(1 - \frac{ak}{h}\right)U_{j+1}^n\right| + \sum_j \left|\left(1 + \frac{ak}{h}\right)U_{j-1}^n\right|\right].$$

But the restriction (10.46) guarantees that

$$1 - \frac{ak}{h} \geq 0, \qquad 1 + \frac{ak}{h} \geq 0$$

and so these can be pulled out of the absolute values, leaving

$$\|U^{n+1}\| \leq \frac{h}{2}\left[\left(1 - \frac{ak}{h}\right)\sum_j |U_{j+1}^n| + \left(1 + \frac{ak}{h}\right)\sum_j |U_{j-1}^n|\right]$$

$$= \frac{1}{2}\left[\left(1 - \frac{ak}{h}\right)\|U^n\| + \left(1 + \frac{ak}{h}\right)\|U^n\|\right]$$

$$= \|U^n\|$$

as desired. This shows that (10.46) is sufficient for stability. In fact, it is also necessary, as we will see in the next section.

Now consider a linear system of equations, and recall that we can decouple the system by making a change of variables to $v(x,t) = R^{-1}u(x,t)$, where R is the matrix of eigenvectors of A. Then v satisfies

$$v_t + \Lambda v_x = 0, \tag{10.47}$$

giving m scalar advection equations.

We can apply this same approach to the difference equations. If we set $V_j^n = R^{-1}U_j^n$ for all j and n and multiply the Lax-Friedrichs method by R^{-1}, we obtain

$$V_j^{n+1} = \frac{1}{2}(V_{j-1}^n + V_{j+1}^n) + \frac{k}{2h}\Lambda(V_{j+1}^n - V_{j-1}^n).$$

This decouples to give m independent difference equations. The pth scalar equation is stable provided

$$\left|\frac{\lambda_p k}{h}\right| \leq 1 \tag{10.48}$$

according to the previous analysis for the scalar case. If (10.48) is satisfied for all $p = 1, 2, \ldots, m$ then V_j^n converges to $v(x, t)$ and hence $U_j^n = RV_j^n$ converges to $u(x, t) = Rv(x, t)$. We see that the Lax-Friedrichs method is stable for a linear system provided that (10.48) is satisfied for each eigenvalue λ_p of A.

EXAMPLE 10.2. Next consider the one-sided method

$$U_j^{n+1} = U_j^n - \frac{ak}{h}(U_j^n - U_{j-1}^n) \tag{10.49}$$

for the scalar equation. Proceeding as in Lax-Friedrichs, we find that

$$\|U^{n+1}\| \leq h \sum_j \left|\left(1 - \frac{ak}{h}\right)U_j^n\right| + h \sum_j \left|\frac{ak}{h}U_{j-1}^n\right|.$$

If we now assume that

$$0 \leq \frac{ak}{h} \leq 1 \tag{10.50}$$

then the coefficients of U_j^n and U_{j-1}^n are both nonnegative, and we obtain the stability bound $\|U^{n+1}\| \leq \|U^n\|$.

Note that the condition (10.50) requires in particular that $a \geq 0$, since k, $h > 0$. The one-sided method (10.49) can only be used when $a \geq 0$. For a system of equations, the same analysis as for Lax-Friedrichs shows that we require

$$0 \leq \frac{\lambda_p k}{h} \leq 1$$

for all eigenvalues of A. In particular, all of the eigenvalues must be nonnegative in order to use this method.

Similarly, the one-sided method using data from the other direction,

$$U_j^{n+1} = U_j^n - \frac{k}{h}A(U_{j+1}^n - U_j^n) \tag{10.51}$$

can be used only if all eigenvalues of A are nonpositive, and the stability restriction is

$$-1 \leq \frac{\lambda_p k}{h} \leq 0. \tag{10.52}$$

The approach used here to investigate the stability of Lax-Friedrichs and one-sided methods works because these methods belong to a special class called "monotone methods". These will be discussed later for nonlinear problems, where a very similar stability analysis can be applied. For methods such as Lax-Wendroff or Leapfrog, which are not monotone methods, a more complicated stability analysis is required. For linear problems this is usually accomplished using Fourier analysis (the von Neumann approach). This very important technique will not be discussed here since it does not generalize to nonlinear problems. (See [63] or [82], for example.) We simply note that the stability restriction for both Lax-Wendroff and Leapfrog agrees with that for Lax-Friedrichs, and is given by (10.48).

10.6   The CFL condition

One of the first papers on finite difference methods for PDEs was written in 1928 by
Courant, Friedrichs, and Lewy[12]. (There is an English translation in [13].) They use
finite difference methods as an analytic tool for proving existence of solutions of certain
PDEs. The idea is to define a sequence of approximate solutions (via finite difference
equations), prove that they converge as the grid is refined, and then show that the limit
function must satisfy the PDE, giving existence of a solution.

In the course of proving convergence of this sequence (which is precisely what we
are interested in), they recognized that a necessary stability condition for any numerical
method is that the domain of dependence of the finite difference method should include
the domain of dependence of the PDE, at least in the limit as k, $h \to 0$. This condition
is known as the CFL condition after Courant, Friedrichs, and Lewy.

The domain of dependence $\mathcal{D}(\bar{x}, \bar{t})$ for the PDE has already been described in Chapter 3. Recall that for the linear system (10.1), the set $\mathcal{D}(\bar{x}, \bar{t})$ consists of the points
$\bar{x} - \lambda_p \bar{t}$, $p = 1, 2, \ldots, m$ since only initial data at these points can affect the solution at
$(\bar{x}, \bar{t})$.

The numerical domain of dependence, $\mathcal{D}_k(\bar{x}, \bar{t})$, for a particular method is similarly
defined. It is the set of points x for which initial data $u_0(x)$ could possibly affect the
numerical solution at $(\bar{x}, \bar{t})$.

EXAMPLE 10.3.   Consider a three-point scheme with a stencil as on the left in
Figure 10.1. The value of $U_k(x_j, t_n)$ depends on values of U_k at time t_{n-1} at the three
points x_{j+q}, $q = -1, 0, 1$. These values in turn depend on U_k at time t_{n-2} at the points
x_{j+q} for $q = -2, -1, 0, 1, 2$. Continuing back to $t = 0$ we see that the solution at t_n
depends only on initial data at the points x_{j+q} for $q = -n, \ldots, n$, and so the numerical
domain of dependence satisfies

$$\mathcal{D}_k(x_j, t_n) \subset \{x : |x - x_j| \leq nh\},$$

or, in terms of a fixed point $(\bar{x}, \bar{t})$,

$$\mathcal{D}_k(\bar{x}, \bar{t}) \subset \{x : |x - \bar{x}| \leq (\bar{t}/k)h\}.$$

If we now refine the mesh with k/h fixed, say $k/h = r$, then in the limit this domain of
dependence fills out the entire set,

$$\mathcal{D}_0(\bar{x}, \bar{t}) = \{x : |x - \bar{x}| \leq \bar{t}/r\}. \tag{10.53}$$

The CFL condition requires that

$$\mathcal{D}(\bar{x}, \bar{t}) \subset \mathcal{D}_0(\bar{x}, \bar{t}). \tag{10.54}$$

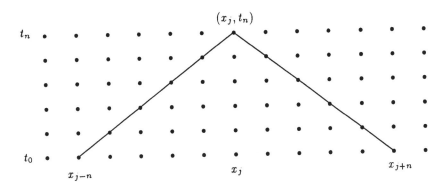

Figure 10.2. Computational triangle for a three-point scheme. The CFL condition requires that characteristics lie within this triangle.

For a three-point scheme applied to a linear system, this requires that

$$|(\bar{x} - \lambda_p \bar{t}) - \bar{x}| \leq \bar{t}/r,$$

i.e., that

$$\left|\frac{\lambda_p k}{h}\right| \leq 1 \tag{10.55}$$

for each eigenvalue λ_p of A. This guarantees that the characteristics of the system lie within the computational triangle shown in Figure 10.2.

The CFL condition is only a necessary condition for stability, not sufficient. The fact that it is necessary stems from the observation that if (10.54) is violated then there are points ξ in the true domain of dependence that are not in the numerical domain of dependence. Changing the value of the initial data at ξ would thus effect the true solution but not the numerical solution, and hence the numerical solution cannot possibly converge to the true solution for all initial data.

For the Lax-Friedrichs method we have already seen that (10.55) is also a sufficient condition for stability, hence it is necessary and sufficient. Similarly, the one-sided method (10.51) has domain of dependence

$$\mathcal{D}_0(\bar{x}, \bar{t}) = \{x : \bar{x} \leq x \leq \bar{x} + \bar{t}/r\},$$

and hence the CFL condition requires precisely the condition (10.52) that we have already shown to be sufficient.

There are examples, however, when the CFL condition is not sufficient for stability. For example, the three-point centered method (10.7) has the usual domain of dependence (10.53) for a three-point method, and hence satisfies the CFL condition if (10.55) holds, but turns out to be unstable for any mesh ratio as is easily shown by von Neumann analysis.

The quantity

$$\nu = \max_p \left| \frac{\lambda_p k}{h} \right| \tag{10.56}$$

is often called the **Courant number**, or sometimes the CFL number, for a particular calculation. A necessary condition for stability of centered three-point methods is thus that the Courant number be no greater than 1.

10.7   Upwind methods

For the scalar advection equation with $a > 0$, the one-sided method (10.49) can be applied and is stable provided (10.50) is satisfied. This method is usually called the **first order upwind method**, since the one-sided stencil points in the "upwind" or "upstream" direction, the correct direction from which characteristic information propagates. If we think of the advection equation as modeling the advection of a concentration profile in a fluid stream, then this is literally the upstream direction.

Similarly, the method (10.51) is the upwind method for the advection equation with $a < 0$.

For a system of equations, we have seen that a one-sided method can only be used if all of the eigenvalues of A have the same sign. This is typically not the case in practice. In the linearized Euler equations, for example, the eigenvalues of $f'(u)$ are v, $v \pm c$ and these all have the same sign only if $|v| > c$. In this case the fluid speed is greater than the sound speed and the flow is said to be **supersonic**. If $|v| < c$ then the flow is **subsonic** and any one-sided method will be unstable since there are sound waves propagating in both directions and the CFL condition is violated.

When computing discontinuous solutions, upwind differencing turns out to be an important tool, even for indefinite systems with eigenvalues of mixed sign. The appropriate application of upwind methods requires some sort of decomposition into characteristic fields. For example, if we change variables and decouple the linear system into (10.47), then each of the resulting scalar problems can be solved with an appropriate upwind method, using the point to the left when $\lambda_p > 0$ or to the right when $\lambda_p < 0$.

This method can be written in a compact form if we introduce the notation

$$\lambda_p^+ = \max(\lambda_p, 0), \qquad \Lambda^+ = \mathrm{diag}(\lambda_1^+, \ldots, \lambda_m^+), \tag{10.57}$$
$$\lambda_p^- = \min(\lambda_p, 0), \qquad \Lambda^- = \mathrm{diag}(\lambda_1^-, \ldots, \lambda_m^-). \tag{10.58}$$

Note that $\Lambda^+ + \Lambda^- = \Lambda$. Then the upwind method for the system (10.47) can be written as

$$V_j^{n+1} = V_j^n - \frac{k}{h}\Lambda^+(V_j^n - V_{j-1}^n) - \frac{k}{h}\Lambda^-(V_{j+1}^n - V_j^n). \tag{10.59}$$

We can transform this back to the original U variables by multiplying by R. This gives

$$U_j^{n+1} = U_j^n - \frac{k}{h}A^+(U_j^n - U_{j-1}^n) - \frac{k}{h}A^-(U_{j+1}^n - U_j^n). \tag{10.60}$$

where

$$A^+ = R\Lambda^+ R^{-1}, \qquad A^- = R\Lambda^- R^{-1}. \tag{10.61}$$

Notice that $A^+ + A^- = A$. If all the eigenvalues of A have the same sign then either A^+ or A^- is zero and the method (10.60) reduces to a fully one-sided method.

For nonlinear systems analogous splittings can be introduced in various ways to incorporate upwinding. Many of these methods require solving Riemann problems in order to accomplish the appropriate splitting between wave propagation to the left and right. The fundamental method of this type is **Godunov's method**. We will see later that (10.60) is precisely Godunov's method applied to a linear system.

11 Computing Discontinuous Solutions

For conservation laws we are naturally interested in the difficulties caused by discontinuities in the solution. In the linear theory presented so far we have assumed smooth solutions, and used this in our discussion of the truncation error and convergence proof. We now consider what happens when we apply a numerical method to a linear problem with discontinuous initial data, e.g., the Riemann problem for the scalar advection equation

$$u_t + au_x = 0, \quad -\infty < x < \infty, \quad t \geq 0,$$
$$u_0(x) = \begin{cases} 1 & x < 0 \\ 0 & x > 0. \end{cases} \tag{11.1}$$

Of course the exact solution is simply $u_0(x - at)$, but we expect our numerical method to have difficulty near the discontinuity. Note for example that a finite difference approximation to u_x applied across the discontinuity in the true solution will blow up as $h \to 0$, e.g.,

$$\frac{u(at + h, t) - u(at - h, t)}{2h} = \frac{0 - 1}{2h} \to -\infty \text{ as } h \to 0.$$

The local truncation error does not vanish as $h \to 0$ and the proof of convergence presented in the previous chapter breaks down. This proof can be rescued by considering approximations $u_0^\epsilon(x)$ to $u_0(x)$ that are smooth and approach u_0 as $\epsilon \to 0$. Although convergence can again be proved for a stable and consistent method, the *rate* of convergence may be considerably less than what is expected from the "order" of the method as defined by its behavior on smooth solutions. Moreover, the numerical solution may look very unsatisfactory on any particular finite grid.

Figures 11.1 and 11.2 show numerical solutions to the Riemann problem (11.1) computed with some of the standard methods discussed in the previous chapter. In all cases $a = 1$, $k/h = 0.5$ and the results are plotted at time $t = 0.5$ along with the exact solution. In Figure 11.1, $h = 0.01$ while a finer grid with $h = 0.0025$ is used in Figure 11.2. Notice

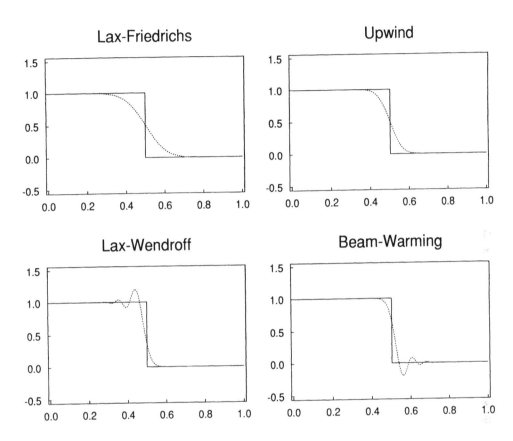

Figure 11.1. Numerical and exact solution to (11.1) with h = 0.01 and the following methods: (a) Lax-Friedrichs, (b) Upwind, (c) Lax-Wendroff, (d) Beam-Warming.

that the first order methods (Lax-Friedrichs and upwind) give very smeared solutions while the second order methods (Lax-Wendroff and Beam-Warming) give oscillations. This behavior is typical.

If we compute the 1-norm error in these computed solutions we do not see the expected rates of convergence. Instead the "first order" methods converge with an error that is $O(k^{1/2})$ while the "second order" methods have an error that is $O(k^{2/3})$ at best. These convergence rates can be proved for very general initial data by a careful analysis using smooth approximations. An indication of why this should be so will be given later in this chapter.

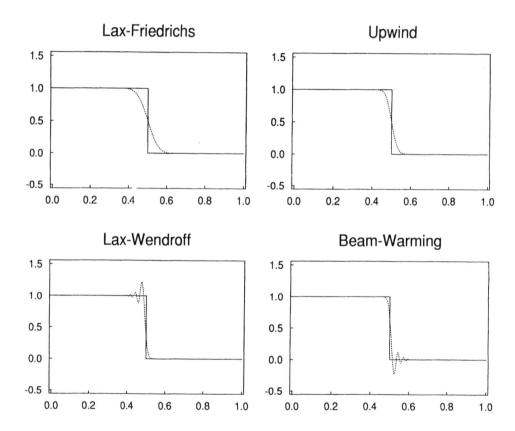

Figure 11.2. Numerical and exact solution to (11.1) with $h = 0.0025$ and the following methods: (a) Lax-Friedrichs, (b) Upwind, (c) Lax-Wendroff, (d) Beam-Warming.

11.1 Modified equations

A useful technique for studying the behavior of solutions to difference equations is to model the difference equation by a differential equation. Of course the difference equation was originally derived by approximating a PDE, and so we can view the original PDE as a model for the difference equation, but there are other differential equations that are better models. In other words, there are other PDEs that the numerical method solves more accurately than the original PDE.

At first glance it may seem strange to approximate the difference equation by a PDE. The difference equation was introduced in the first place because it is easier to solve than the PDE. This is true if we want to generate numerical approximations, but on the other hand it is often easier to predict the *qualitative* behavior of a PDE than of a system of difference equations. At the moment it is the qualitative behavior of the numerical methods that we wish to understand. Good descriptions of the theory and use of modified equations can be found in Hedstrom[37] or Warming-Hyett[96].

The derivation of the modified equation is closely related to the calculation of the local truncation error for a given method. Consider Lax-Friedrichs, for example, where we calculated the local truncation error $L_k(x,t)$ from (10.29). Since $u(x,t)$ is taken to be the exact solution to $u_t + Au_x = 0$, we found that $L_k(x,t) = O(k)$. However, if we instead take $u(x,t)$ to be the solution of the PDE

$$u_t + Au_x + \frac{1}{2}\left(ku_{tt} - \frac{h^2}{k}u_{xx}\right) = 0, \tag{11.2}$$

then the truncation error would be $O(k^2)$. We conclude that the Lax-Friedrichs method produces a *second order* accurate approximation to the solution of (11.2). This equation is called a **modified equation** for the Lax-Friedrichs method. (The term "model equation" is also sometimes used.)

If we express the u_{tt} term in (11.2) in terms of x-derivatives we obtain an equation that is easier to analyze. Note that we cannot use (10.11) directly since u no longer satisfies the original PDE, but a similar manipulation using (11.2) shows that

$$\begin{aligned} u_{tt} &= -Au_{tx} - \frac{1}{2}\left(ku_{ttt} - \frac{h^2}{k}u_{xxt}\right) \\ &= -A[-Au_{xx} + O(k)] + O(k) \\ &= A^2 u_{xx} + O(k). \end{aligned}$$

Using $u_{tt} = A^2 u_{xx}$ in (11.2) will give errors that are $O(k^2)$, the same order as other terms that have already been ignored, and consequently the Lax-Friedrichs method is also second order accurate on the modified equation

$$u_t + Au_x = \frac{h^2}{2k}\left(I - \frac{k^2}{h^2}A^2\right)u_{xx} \tag{11.3}$$

where I is the $m \times m$ identity matrix. For higher order methods this elimination of t-derivatives in terms of x-derivatives can also be done, but must be done carefully and is complicated by the need to include higher order terms. Warming and Hyett[96] present a general procedure.

11.1.1 First order methods and diffusion

The modified equation (11.3) is an advection-diffusion equation of the form

$$u_t + Au_x = Du_{xx} \tag{11.4}$$

with a diffusion matrix D given by

$$D = \frac{h^2}{2k}\left(I - \left(\frac{k}{h}A\right)^2\right). \tag{11.5}$$

We expect solutions of this equation to become smeared out as time evolves, explaining at least the qualitative behavior of the Lax-Friedrichs method seen in the above figures. In fact this equation is even a good quantitative model for how the solution behaves. If we plot the exact solution to (11.3) (with appropriate k and h) along with the Lax-Friedrichs numerical solutions shown in Figures 11.1 and 11.2, they are virtually indistinguishable to plotting accuracy.

Note that the modified equation (11.3) varies with k and h. The diffusive term is $O(k)$ as $k \to 0$ and vanishes in the limit as the grid is refined. The numerical solutions generated by the Lax-Friedrichs method on a sequence of grids are thus good approximations to a sequence of "vanishing viscosity" solutions u^ϵ that might be used to define the physically relevant weak solution to the conservation law. In the linear case there is only one weak solution, but for nonlinear problems it turns out that the Lax-Friedrichs method satisfies a discrete entropy condition and converges more generally to the vanishing viscosity weak solution as the grid is refined.

The modified equation for the upwind method can be derived similarly, and is found to be

$$u_t + Au_x = \frac{1}{2}hA\left(I - \frac{k}{h}A\right)u_{xx}. \tag{11.6}$$

This is again an advection-diffusion equation and so we expect similar behavior. Moreover, for the values of k/h and A used in the computations presented above, the diffusion coefficient is $h/4$ for upwind and $3h/4$ for Lax-Friedrichs, so we expect upwind to be less diffusive than Lax-Friedrichs, as confirmed in the numerical results.

Relation to stability. Notice that the equation (11.3) is mathematically well posed only if the diffusion coefficient matrix D is positive semi-definite. Otherwise it behaves like the backward heat equation which is notoriously ill posed. This requires that the eigenvalues of D be nonnegative. The matrix R of eigenvectors of A also diagonalizes D and we see that the eigenvalues of D are

$$\frac{h^2}{2k}\left(1 - \left(\frac{k\lambda_p}{h}\right)^2\right).$$

These are all nonnegative if and only if the stability condition (10.48) is satisfied. We see that the modified equation also gives some indication of the stability properties of the method.

Similarly, the diffusion matrix for the upwind method has nonnegative eigenvalues if and only if the stability condition $0 \leq \lambda_p k/h \leq 1$ is satisfied.

EXERCISE 11.1. *Compute the modified equation for the method*

$$U_j^{n+1} = U_j^n - \frac{k}{2h}A(U_{j+1}^n - U_{j-1}^n).$$

Explain why this method might be expected to be unstable for all k/h (as in fact it is).

11.1.2   Second order methods and dispersion

The Lax-Wendroff method, which is second order accurate on $u_t + Au_x = 0$, gives a third order accurate approximation to the solution of the modified equation

$$u_t + Au_x = \frac{h^2}{6}A\left(\frac{k^2}{h^2}A^2 - I\right)u_{xxx}. \tag{11.7}$$

The Beam-Warming method has a similar modified equation,

$$u_t + Au_x = \frac{h^2}{6}A\left(2I - \frac{3k}{h}A + \frac{k^2}{h^2}A^2\right)u_{xxx}. \tag{11.8}$$

In the scalar case, both of these modified equations have the form

$$u_t + au_x = \mu u_{xxx} \tag{11.9}$$

which is a **dispersive** equation. The theory of dispersive waves is covered in detail in Whitham[97], for example. The key observation is that if we look at a Fourier series solution to this equation, taking $u(x,t)$ of the form

$$u(x,t) = \int_{-\infty}^{\infty} \hat{u}(\xi, t)e^{i\xi x}\,d\xi, \tag{11.10}$$

then the Fourier components with different wave number ξ propagate at different speeds, *i.e.*, they disperse as time evolves. By linearity it is sufficient to consider each wave number in isolation, so suppose we look for solutions to (11.9) of the form

$$e^{i(\xi x - c(\xi)t)}. \tag{11.11}$$

Putting this into (11.9) and canceling common terms gives

$$c(\xi) = a\xi + \mu\xi^3. \tag{11.12}$$

This expression is called the **dispersion relation** for (11.9). The speed at which this oscillating wave propagates is clearly $c(\xi)/\xi$, which is called the **phase velocity** $c_p(\xi)$ for wave number ξ. This is the speed at which wave peaks travel. From (11.12) we find that

$$c_p(\xi) = c(\xi)/\xi = a + \mu\xi^2. \tag{11.13}$$

Note that this varies with ξ and is close to the propagation speed a of the original advection equation only for ξ sufficiently small.

It turns out that for general data composed of many wavenumbers, a more important velocity is the so-called **group velocity**, defined by

$$c_g(\xi) = c'(\xi) = a + 3\mu\xi^2. \tag{11.14}$$

This varies even more substantially with ξ than $c_p(\xi)$. The importance of the group velocity is discussed in Whitham[97]. See also Brillouin[4] or Lighthill[52]. The utility of this concept in the study of numerical methods has been stressed by Trefethen, in particular in relation to the stability of boundary conditions. A nice summary of some of this theory may be found in Trefethen[86].

A step function, such as the initial data we use in (11.1), has a broad Fourier spectrum: $\hat{u}(\xi, 0)$ decays only like $1/\xi$ as $|\xi| \to 0$. (By contrast, a C^∞ function has an exponentially decaying Fourier transform.) As time evolves these highly oscillatory components disperse, leading to an oscillatory solution as has already been observed in the numerical solution obtained using Lax-Wendroff or Beam-Warming. It can be shown that at time t, wavenumber ξ will be predominantly visible near $x = c_g(\xi)t$. So, according to (11.14), the most oscillatory components are found farthest from the correct location $x = at$. This can also be seen in Figures 11.1 and 11.2.

For the scalar advection equation, the modified equation (11.7) for Lax-Wendroff is of the form (11.9) with

$$\mu = \frac{1}{6}h^2 a(\nu^2 - 1) \tag{11.15}$$

where $\nu = ak/h$ is the Courant number. Since $a > 0$ and $|\nu| < 1$ for stability, we have $\mu < 0$ and hence $c_g(\xi) < a$ for all ξ according to (11.14). All wave numbers travel too slowly, leading to an oscillatory wave train lagging behind the discontinuity in the true solution, as seen in the figures.

For Beam-Warming, on the other hand,

$$\mu = \frac{1}{6}h^2 a(2 - 3\nu + \nu^2) \tag{11.16}$$

which is easily seen to be positive for $0 \le \nu \le 1$. Consequently $c_g(\xi) > a$ for all ξ and the oscillations are ahead of the discontinuity.

11.2 Accuracy

We now return to the question of the accuracy of a numerical method when applied to a problem with discontinuities. As an example we will consider the Lax-Friedrichs method applied to the problem (11.1). Since the numerical solution agrees so well with the true solution of the modified equation, we can use the difference between the true solution of the modified equation and the true solution to the advection equation as an estimate of the error in the numerical approximation. This is not a rigorous error estimate, and is only for the particular initial data of (11.1) anyway, but it does give an accurate indication of what can be expected in general. The connection to the modified equation helps explain why we typically lose accuracy near discontinuities.

The true solution to (11.4) with data $u_0(x)$ from (11.1) is simply

$$u^D(x,t) = 1 - \operatorname{erf}\left((x - at)/\sqrt{4Dt}\right) \tag{11.17}$$

where the "error function" erf is defined by

$$\operatorname{erf}(x) = \frac{2}{\sqrt{\pi}} \int_{-\infty}^{x} e^{-z^2}\, dz. \tag{11.18}$$

From this and the fact that the solution to the pure advection problem is simply $u(x,t) = u_0(x - at)$, we can easily compute that

$$
\begin{aligned}
\|u(\cdot,t) - u^D(\cdot,t)\| &= 2\int_{-\infty}^{0} \operatorname{erf}\left(x/\sqrt{4Dt}\right)\, dx \\
&= 2\sqrt{4Dt} \int_{-\infty}^{0} \operatorname{erf}(z)\, dz \\
&= C_1\sqrt{Dt}
\end{aligned}
$$

for some constant C_1 independent of D and t. For Lax-Friedrichs, D is given by (11.5) and so we find that

$$\|u(\cdot,t) - U_k(\cdot,t)\| \approx C_2\sqrt{ht} \tag{11.19}$$

as $h \to 0$, for k/h fixed. This indicates that the 1-norm of the error decays only like $h^{1/2}$ even though the method is formally "first order accurate".

Similar results for nonlinear scalar conservation laws can be found in Kuznetsov[43], Lucier[54], or Sanders[70], for example.

12 Conservative Methods for Nonlinear Problems

When we attempt to solve nonlinear conservation laws numerically we run into additional difficulties not seen in the linear equation. Moreover, the nonlinearity makes everything harder to analyze. In spite of this, a great deal of progress has been made in recent years.

For *smooth* solutions to nonlinear problems, the numerical method can often be linearized and results from the theory of linear finite difference methods applied to obtain convergence results for nonlinear problems. A very general theorem of this form is due to Strang[80] (see also §5.6 of [63]). We will not pursue this topic here since we are primarily interested in discontinuous solutions, for which very reasonable looking finite difference methods can easily give disasterous results that are obviously (or sometimes, not so obviously) incorrect.

We have already seen some of the difficulties caused by discontinuous solutions even in the linear case. For nonlinear problems there are additional difficulties that can arise:

- The method might be "nonlinearly unstable", *i.e.*, unstable on the nonlinear problem even though linearized versions appear to be stable. Often oscillations will trigger nonlinear instabilities.

- The method might converge to a function that is not a weak solution of our original equation (or that is the wrong weak solution, *i.e.*, does not satisfy the entropy condition).

The fact that we might converge to the wrong weak solution is not so surprising – if there is more than one weak solution why should we necessarily converge to the right one? (We must make sure that the finite difference approximations satisfy some discrete form of the entropy condition, as we will do later.)

The fact that we might converge to a function that is not a weak solution at all is more puzzling, but goes back to the fact that it is possible to derive a variety of conservation laws that are equivalent for smooth solutions but have different weak solutions. For example, the PDEs

$$u_t + \left(\frac{1}{2}u^2\right)_x = 0 \tag{12.1}$$

and

$$(u^2)_t + \left(\frac{2}{3}u^3\right)_x = 0 \tag{12.2}$$

have exactly the same smooth solutions, but the Rankine-Hugoniot condition gives different shock speeds, and hence different weak solutions. (Recall Example 3.3.)

Consider a finite difference method that is consistent with one of these equations, say (12.1), using the same definition of consistency as for linear problems (expand in Taylor series, etc.). Then the method is also consistent with (12.2) since the Taylor series expansion (which assumes smoothness) gives the same result in either case. So the method is consistent with both (12.1) and (12.2) and while we might then expect the method to converge to a function that is a weak solution of both, that is impossible when the two weak solutions differ.

EXAMPLE 12.1. If we write Burgers' equation (12.1) in the quasilinear form

$$u_t + uu_x = 0 \tag{12.3}$$

then a natural finite difference method, obtained by a minor modification of the upwind method for $u_t + au_x = 0$ (and assuming $U_j^n \geq 0$ for all j, n) is

$$U_j^{n+1} = U_j^n - \frac{k}{h}U_j^n(U_j^n - U_{j-1}^n). \tag{12.4}$$

The method (12.4) is adequate for smooth solutions but will not, in general, converge to a discontinuous weak solution of Burgers' equation (12.1) as the grid is refined. Consider, for example, the data from (11.1), which in discrete form gives

$$U_j^0 = \begin{cases} 1 & j < 0 \\ 0 & j \geq 0. \end{cases} \tag{12.5}$$

Then it is easy to verify from (12.4) that $U_j^1 = U_j^0$ for all j. This happens in every successive step as well and so $U_j^n = U_j^0$ for all j and n, regardless of the step sizes k and h. As the grid is refined, the numerical solution thus converges very nicely to the function $u(x,t) = u_0(x)$. This is not a weak solution of (12.1) (or of (12.2) either).

In this example the solution is obviously wrong, but similar behavior is seen with other initial data that may give reasonable looking results that are incorrect. Figure 12.1 shows the true and computed solutions at time $t = 1$ with Riemann data $u_l = 1.2$ and $u_r = 0.4$. We get a nice looking solution propagating at entirely the wrong speed.

EXERCISE 12.1. *Show that (12.4) is consistent with both (12.1) and (12.2).*

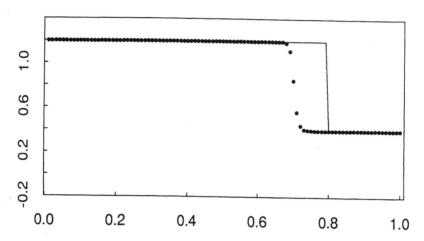

Figure 12.1. True and computed solutions to Burgers' equation using a nonconservative method.

12.1   Conservative methods

Luckily, there turns out to be a very simple and natural requirement we can impose on our numerical methods which will guarantee that we do not converge to non-solutions. This is the requirement that the method be in **conservation form**, which means it has the form

$$U_j^{n+1} = U_j^n - \frac{k}{h}[F(U_{j-p}^n, U_{j-p+1}^n, \ldots, U_{j+q}^n) - F(U_{j-p-1}^n, U_{j-p}^n, \ldots, U_{j+q-1}^n)] \qquad (12.6)$$

for some function F of $p + q + 1$ arguments. F is called the **numerical flux function**. In the simplest case, $p = 0$ and $q = 1$ so that F is a function of only two variables and (12.6) becomes

$$U_j^{n+1} = U_j^n - \frac{k}{h}[F(U_j^n, U_{j+1}^n) - F(U_{j-1}^n, U_j^n)]. \qquad (12.7)$$

This form is very natural if we view U_j^n as an approximation to the cell average $\bar{u}_j^n$ defined by (10.3). We know that the weak solution $u(x, t)$ satisfies the integral form of the conservation law,

$$\int_{x_{j-1/2}}^{x_{j+1/2}} u(x, t_{n+1})\, dx = \int_{x_{j-1/2}}^{x_{j+1/2}} u(x, t_n)\, dx \qquad (12.8)$$

$$- \left[\int_{t_n}^{t_{n+1}} f(u(x_{j+1/2}, t))\, dt - \int_{t_n}^{t_{n+1}} f(u(x_{j-1/2}, t))\, dt \right]$$

Dividing by h and using the cell averages defined in (10.3), this gives

$$\bar{u}_j^{n+1} = \bar{u}_j^n - \frac{1}{h} \left[\int_{t_n}^{t_{n+1}} f(u(x_{j+1/2}, t))\, dt - \int_{t_n}^{t_{n+1}} f(u(x_{j-1/2}, t))\, dt \right] \qquad (12.9)$$

Comparing this to (12.7), we see that the numerical flux function $F(U_j, U_{j+1})$ plays the role of an average flux through $x_{j+1/2}$ over the time interval $[t_n, t_{n+1}]$:

$$F(U_j, U_{j+1}) \sim \frac{1}{k} \int_{t_n}^{t_{n+1}} f(u(x_{j+1/2}, t)) \, dt. \tag{12.10}$$

One way to derive numerical methods in conservation form is to use standard finite difference discretizations but to start with the conservative form of the PDE rather than the quasilinear form. For example, if we generalize the upwind method to Burgers' equation using the form (12.1) rather than (12.3), we obtain

$$U_j^{n+1} = U_j^n - \frac{k}{h} \left[\frac{1}{2}(U_j^n)^2 - \frac{1}{2}(U_{j-1}^n)^2 \right]. \tag{12.11}$$

This is of the form (12.7) with

$$F(v, w) = \frac{1}{2} v^2. \tag{12.12}$$

Here we again assume that $U_j^n \geq 0$ for all j, n, so that the "upwind" direction is always to the left. More generally, for a nonlinear system $u_t + f(u)_x = 0$ for which the Jacobian matrix $f'(U_j^n)$ has nonnegative eigenvalues for all U_j^n, the upwind method is of the form (12.7) with

$$F(v, w) = f(v). \tag{12.13}$$

Of course in general the Jacobian will have eigenvalues of mixed sign and it will not be possible to use a completely one-sided method. Also note that for the nonlinear problem, unlike the linear advection equation, the "upwind" direction depends on the data U_j^n and may vary from point to point. Even in the scalar case we need to introduce some way of switching the directional bias based on the data. We will consider various generalizations of (12.13) to handle this situation later. For now we simply note that if $f'(U_j^n)$ has only nonpositive eigenvalues for all U_j^n, then the upwind method always uses the point to the right, and the flux becomes

$$F(v, w) = f(w). \tag{12.14}$$

Lax-Friedrichs. The generalization of the Lax-Friedrichs method to nonlinear systems takes the form

$$U_j^{n+1} = \frac{1}{2}(U_{j-1}^n + U_{j+1}^n) - \frac{k}{2h} \left(f(U_{j+1}^n) - f(U_{j-1}^n) \right). \tag{12.15}$$

This method can be written in the conservation form (12.7) by taking

$$F(U_j, U_{j+1}) = \frac{h}{2k}(U_j - U_{j+1}) + \frac{1}{2}(f(U_j) + f(U_{j+1})). \tag{12.16}$$

12.2   Consistency

The method (12.7) is consistent with the original conservation law if the numerical flux function F reduces to the true flux f for the case of constant flow. If $u(x,t) \equiv \bar{u}$, say, then by the correspondence (12.10) we expect

$$F(\bar{u}, \bar{u}) = f(\bar{u}) \qquad \forall \bar{u} \in \mathbb{R}. \tag{12.17}$$

Some smoothness is also required, so that as the two arguments of F approach some common value $\bar{u}$, the value of F approaches $f(\bar{u})$ smoothly. For consistency it suffices to have F a **Lipschitz continuous** function of each variable. We say that F is Lipschitz at $\bar{u}$ if there is a constant $K \geq 0$ (which may depend on $\bar{u}$) such that

$$|F(v, w) - f(\bar{u})| \leq K \max(|v - \bar{u}|, |w - \bar{u}|) \tag{12.18}$$

for all v, w with $|v - \bar{u}|$ and $|w - \bar{u}|$ sufficiently small. We say that F is a Lipschitz continuous function if it is Lipschitz at every point.

EXAMPLE 12.2.   The upwind flux (12.13) is consistent since it clearly satisfies (12.17) and is Lipschitz continuous provided f is Lipschitz. Since we are always assuming f is smooth, this will be the case (Note that any differentiable function is Lipschitz).

EXERCISE 12.2.   *Verify that the Lax-Friedrichs flux (12.15) is consistent (including Lipschitz continuity).*

More generally, if the flux F depends on more than two arguments, as in (12.6), the method is consistent if $F(\bar{u}, \bar{u}, \ldots, \bar{u}) = f(\bar{u})$ and the Lipschitz condition requires the existence of a constant K such that

$$|F(U_{j-p}, \ldots, U_{j+q}) - f(\bar{u})| \leq K \max_{-p \leq i \leq q} |U_{j+i} - \bar{u}| \tag{12.19}$$

for all U_{j+i} sufficiently close to $\bar{u}$.

Notation. It is often convenient to write (12.6) in the form

$$U_j^{n+1} = U_j^n - \frac{k}{h}[F(U^n; j) - F(U^n; j-1)] \tag{12.20}$$

where $F(U^n; j)$ is the flux function which is allowed to depend on any (finite) number of elements of the vector U^n, "centered" about the jth point. For example, the upwind flux (12.13) would simply take the form

$$F(U^n; j) = f(U_j^n). \tag{12.21}$$

In this notation, we can think of $F(U^n; j)$ as approximating the average flux

$$F(U^n; j) \sim \frac{1}{k} \int_{t_n}^{t_{n+1}} f(u(x_{j+1/2}, t)) \, dt. \tag{12.22}$$

The notation $F(U^n; j)$ is consistent with the notation $U_j^{n+1} = \mathcal{H}_k(U^n; j)$ introduced in Chapter 10, and for a method in conservation form we have

$$\mathcal{H}_k(U^n; j) = U_j^n - \frac{k}{h}[F(U^n; j) - F(U^n; j - 1)]. \tag{12.23}$$

Lax-Wendroff and MacCormack methods. Recall that the Lax-Wendroff method for a constant coefficient linear hyperbolic system $u_t + Au_x = 0$ has the form

$$U_j^{n+1} = U_j^n - \frac{k}{2h}A(U_{j+1}^n - U_{j-1}^n) + \frac{k^2}{2h^2}A^2(U_{j+1}^n - 2U_j^n + U_{j-1}^n). \tag{12.24}$$

There are various ways that this can be extended to give a second order method for nonlinear conservation laws. If we let $A(u) = f'(u)$ be the Jacobian matrix, then a conservative form of Lax-Wendroff is

$$\begin{aligned}
U_j^{n+1} = \; & U_j^n - \frac{k}{2h}(f(U_{j+1}^n) - f(U_{j-1}^n)) + \frac{k^2}{2h^2}\Big[A_{j+1/2}(f(U_{j+1}^n) - f(U_j^n)) \\
& - A_{j-1/2}(f(U_j^n) - f(U_{j-1}^n))\Big],
\end{aligned} \tag{12.25}$$

where $A_{j\pm1/2}$ is the Jacobian matrix evaluated at $\frac{1}{2}(U_j^n + U_{j\pm1}^n)$. The difficulty with this form is that it requires evaluating the Jacobian matrix, and is more expensive to use than other forms that only use the function $f(u)$.

One way to avoid using A is to use a two-step procedure. This was first proposed by Richtmyer, and the **Richtmyer two-step Lax-Wendroff method** is

$$\begin{aligned}
U_{j+1/2}^{n+1/2} &= \frac{1}{2}(U_j^n + U_{j+1}^n) - \frac{k}{2h}\left[f(U_{j+1}^n) - f(U_j^n)\right] \\
U_j^{n+1} &= U_j^n - \frac{k}{h}\left[f(U_{j+1/2}^{n+1/2}) - f(U_{j-1/2}^{n+1/2})\right].
\end{aligned} \tag{12.26}$$

Another method of this same type was proposed by MacCormack[55]. **MacCormack's method** uses first forward differencing and then backward differencing to achieve second order accuracy:

$$\begin{aligned}
U_j^* &= U_j^n - \frac{k}{h}\left[f(U_{j+1}^n) - f(U_j^n)\right] \\
U_j^{n+1} &= \frac{1}{2}(U_j^n + U_j^*) - \frac{k}{2h}\left[f(U_j^*) - f(U_{j-1}^*)\right].
\end{aligned} \tag{12.27}$$

Alternatively, we could use backward differencing in the first step and then forward differencing in the second step.

EXERCISE 12.3. *Each of the methods (12.25), (12.26) and (12.27) reduces to (12.24) in the constant coefficient linear case and is second order accurate on smooth solutions (to nonlinear problems) and conservative. Verify these statements for at least one of these methods and write it in conservation form, determining the numerical flux function.*

12.3   Discrete conservation

The basic principle underlying a conservation law is that the total quantity of a conserved variable in any region changes only due to flux through the boundaries. This gave the integral form of the conservation law,

$$\int_a^b u(x, t_2)\, dx = \int_a^b u(x, t_1)\, dx - \left(\int_{t_1}^{t_2} f(u(b, t))\, dt - \int_{t_1}^{t_2} f(u(a, t))\, dt \right), \qquad (12.28)$$

which holds for any a, b, t_1 and t_2. Notice in particular that if u is identically constant outside some finite interval over the time interval $t_1 \le t \le t_2$, say $u \equiv u_{-\infty}$ for $x \le a$ and $u \equiv u_{+\infty}$ for $x \ge b$, then we obtain

$$\int_a^b u(x, t_2)\, dx = \int_a^b u(x, t_1)\, dx - (t_2 - t_1)(f(u_{+\infty}) - f(u_{-\infty})). \qquad (12.29)$$

(Note that by finite propagation speed this will be the case if the initial data is constant outside some finite interval.) If $u_{+\infty} = u_{-\infty}$ (e.g., if the data has compact support in which case $u_{\pm\infty} = 0$), then the flux terms drop out altogether and the integral $\int_a^b u(x, t)\, dx$ is constant in time over any time interval for which the solution remains constant at a and b.

We have already used (12.28) to motivate the conservation form (12.6) by applying this with $a = x_{j-1/2}$, $b = x_{j+1/2}$. But we can easily show that a numerical solution generated by a conservative method will also have a more global form of conservation, analogous to (12.28) for arbitrary a, b. In the discrete case, if we let $J < K$ be arbitrary cell indices and sum (12.20) over j, we find that

$$h \sum_{j=J}^{K} U_j^{n+1} = h \sum_{j=J}^{K} U_j^n - k \sum_{j=J}^{K} [F(U^n; j) - F(U^n; j - 1)]. \qquad (12.30)$$

The last sum here telescopes, and all fluxes drop out except for the fluxes at the extreme cell boundaries $x_{J-1/2}$ and $x_{K+1/2}$, so that (12.30) reduces to a discrete form of (12.28) over the time interval $[t_n, t_{n+1}]$:

$$h \sum_{j=J}^{K} U_j^{n+1} = h \sum_{j=J}^{K} U_j^n - k[F(U^n; K) - F(U^n; J - 1)]. \qquad (12.31)$$

In particular, if u_0 is constant outside some finite interval then so is U^n, since explicit numerical methods have finite domain of dependence, and so for J and K sufficiently far out, we can use consistency of the flux function F to obtain

$$h \sum_{j=J}^{K} U_j^{n+1} = h \sum_{j=J}^{K} U_j^n - k[f(u_{+\infty}) - f(u_{-\infty})]. \qquad (12.32)$$

Applying this recursively gives, for $N > n$,

$$h \sum_{j=J}^{K} U_j^N = h \sum_{j=J}^{K} U_j^n - (t_N - t_n)[f(u_{+\infty}) - f(u_{-\infty})], \qquad (12.33)$$

a discrete form of (12.29).

It follows that if

$$h \sum_{j=J}^{K} U_j^0 = \int_{x_{J-1/2}}^{x_{K+1/2}} u_0(x) \, dx, \qquad (12.34)$$

which will hold for example if we take $U_j^0 = \bar{u}_j^0$, the cell averages, then we also have

$$h \sum_{j=J}^{K} U_j^n = \int_{x_{J-1/2}}^{x_{K+1/2}} u(x, t_n) \, dx, \qquad (12.35)$$

for all n small enough that the solution remains constant in the neighborhood of $x_{J-1/2}$ and $x_{K+1/2}$. Using the notation $U_k(x, t)$ for the piecewise constant function defined by U_j^n, we have

$$\int_{x_{J-1/2}}^{x_{K+1/2}} U_k(x, t_n) \, dx = \int_{x_{J-1/2}}^{x_{K+1/2}} u(x, t_n) \, dx, \qquad (12.36)$$

and so we say that the discrete method is **conservative**.

This discrete conservation means that any shocks we compute must, in a sense, be in the "correct" location. Consider, for example, the test problem for Burgers' equation with data (12.5). We have seen in Example 12.1 that a nonconservative method can give a solution with the shock propagating at the wrong speed. This could not happen with a conservative method, since the integral of $U_k(x, t)$ is obviously increasing at the wrong rate. The solution computed with a conservative method might have the shock smeared out, but since the integral (12.36) is correct, it must at least be smeared about the correct location.

Figure 12.2 shows the same test case as seen in Figure 12.1, but now using the conservative upwind method (12.11). Note that the total integral of $U_k - u$ appears to be zero, as expected.

12.4   The Lax-Wendroff Theorem

The above discussion suggests that we can hope to correctly approximate discontinuous weak solutions to the conservation law by using a conservative method.

Lax and Wendroff[46] proved that this is true, at least in the sense that if we converge to some function $u(x, t)$ as the grid is refined, through some sequence k_l, $h_l \to 0$, then this function will in fact be a weak solution of the conservation law. The theorem does not guarantee that we do converge. For that we need some form of stability, and even then

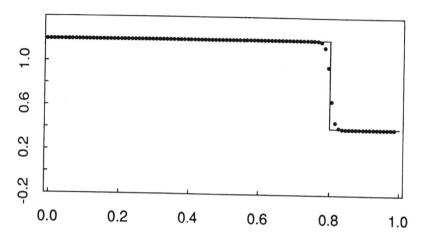

Figure 12.2. True and computed solutions to Burgers' equation using the conservative upwind method.

if there is more than one weak solution it might be that one sequence of approximations will converge to one weak solution, while another sequence converges to a different weak solution (and therefore a third sequence, obtained for example by merging together the first two sequences, will not converge at all!). See Exercise 12.4 below for an example of this.

Nonetheless, this is a very powerful and important theorem, for it says that we can have confidence in solutions we compute. In practice we typically do not consider a whole sequence of approximations. Instead we compute a single approximation on one fixed grid. If this solution looks reasonable and has well-resolved discontinuities (an indication that the method is stable and our grid is sufficiently fine), then we can believe that it is in fact a good approximation to some weak solution.

THEOREM 12.1 (LAX AND WENDROFF[46]). *Consider a sequence of grids indexed by $l = 1, 2, \ldots$, with mesh parameters k_l, $h_l \to 0$ as $l \to \infty$. Let $U_l(x, t)$ denote the numerical approximation computed with a consistent and conservative method on the lth grid. Suppose that U_l converges to a function u as $l \to \infty$, in the sense made precise below. Then $u(x, t)$ is a weak solution of the conservation law.*

We will assume that we have convergence of U_l to u in the following sense:

1. Over every bounded set $\Omega = [a, b] \times [0, T]$ in x-t space,

$$\int_0^T \int_a^b |U_l(x, t) - u(x, t)| \, dx \, dt \to 0 \quad \text{as } l \to \infty. \tag{12.37}$$

This is the 1-norm over the set Ω, so we can simply write

$$\|U_l - u\|_{1,\Omega} \to 0 \quad \text{as } l \to \infty. \tag{12.38}$$

2. We also assume that for each T there is an $R > 0$ such that

$$TV(U_l(\cdot, t)) < R \quad \text{for all } 0 \le t \le T, \quad l = 1, 2, \ldots. \tag{12.39}$$

Here TV denotes the total variation function,

$$TV(v) = \sup \sum_{j=1}^{N} |v(\xi_j) - v(\xi_{j-1})| \tag{12.40}$$

where the supremum is taken over all subdivisions of the real line $-\infty = \xi_0 < \xi_1 < \cdots < \xi_N = \infty$. Note that for the total variation to be finite v must approach constant values $v_{\pm\infty}$ as $x \to \pm\infty$.

Another possible definition is

$$TV(v) = \limsup_{\epsilon \to 0} \frac{1}{\epsilon} \int_{-\infty}^{\infty} |v(x) - v(x - \epsilon)| \, dx. \tag{12.41}$$

If v is differentiable then this reduces to

$$TV(v) = \int_{-\infty}^{\infty} |v'(x)| \, dx. \tag{12.42}$$

We can use (12.42) also for nondifferentiable functions (distributions) if we interpret $v'(x)$ as the distribution derivative (which includes delta functions at points where v is discontinuous).

Lax and Wendroff assumed a slightly different form of convergence, namely that U_l converges to u almost everywhere (*i.e.*, except on a set of measure zero) in a uniformly bounded manner. Using the fact that each U_l is a piecewise constant function, it can be shown that this requirement is essentially equivalent to (12.38) and (12.39) above. The advantage of assuming (12.38) and (12.39) is twofold: *a)* it is these properties that are really needed in the proof, and *b)* for certain important classes of methods (*e.g.*, the "total variation diminishing" methods), it is this form of convergence that we can most directly prove.

PROOF. We will show that the limit function $u(x, t)$ satisfies the weak form (3.22), *i.e.*, for all $\phi \in C_0^1$,

$$\int_0^\infty \int_{-\infty}^{+\infty} [\phi_t u + \phi_x f(u)] \, dx \, dt = - \int_{-\infty}^\infty \phi(x, 0) u(x, 0) \, dx. \tag{12.43}$$

Let ϕ be a C_0^1 test function and multiply the numerical method (12.20) by $\phi(x_j, t_n)$, yielding

$$\phi(x_j, t_n)U_j^{n+1} = \phi(x_j, t_n)U_j^n - \frac{k}{h}\phi(x_j, t_n)[F(U^n; j) - F(U^n; j-1)].\qquad(12.44)$$

This is true for all values of j and n on each grid l. (Dropping the subscript l on k and h and letting U_j^n represent the pointwise values of U_l in order to make the formulas more readable.)

If we now sum (12.44) over all j and $n \geq 0$, we obtain

$$\sum_{n=0}^{\infty}\sum_{j=-\infty}^{\infty}\phi(x_j, t_n)(U_j^{n+1} - U_j^n)\qquad(12.45)$$

$$= \frac{k}{h}\sum_{n=0}^{\infty}\sum_{j=-\infty}^{\infty}\phi(x_j, t_n)[F(U^n; j) - F(U^n; j-1)].$$

We now use "summation by parts", which just amounts to recombining the terms in each sum. A simple example is

$$\sum_{j=1}^{m}a_j(b_j - b_{j-1}) = (a_1 b_1 - a_1 b_0) + (a_2 b_2 - a_2 b_1) + \cdots + (a_m b_m - a_m b_{m-1})$$

$$= -a_1 b_0 + (a_1 b_1 - a_2 b_1) + (a_2 b_2 - a_3 b_2) +$$

$$\cdots + (a_{m-1}b_{m-1} - a_m b_{m-1}) + a_m b_m\qquad(12.46)$$

$$= a_m b_m - a_1 b_0 - \sum_{j=1}^{m-1}(a_{j+1} - a_j)b_j.$$

Note that the original sum involved the product of a_j with differences of b's whereas the final sum involves the product of b_j with differences of a's. This is completely analogous to integration by parts, where the derivative is moved from one function to the other. Just as in integration by parts, there are also boundary terms $a_m b_m - a_1 b_0$ that arise.

We will use this on each term in (12.45) (on the n-sum in the first term and on the j-sum in the second term). By our assumption that ϕ has compact support, $\phi(x_j, t_n) = 0$ for $|j|$ or n sufficiently large, and hence the boundary terms at $j = \pm\infty$, $n = \infty$ all drop out. The only boundary term that remains is at $n = 0$, where $t_0 = 0$. This gives

$$-\sum_{j=-\infty}^{\infty}\phi(x_j, t_0)U_j^0 - \sum_{n=1}^{\infty}(\phi(x_j, t_n) - \phi(x_j, t_{n-1}))U_j^n$$

$$-\frac{k}{h}\sum_{n=0}^{\infty}\sum_{j=-\infty}^{\infty}(\phi(x_{j+1}, t_n) - \phi(x_j, t_n))F(U^n; j) = 0.$$

Note that each of these sums is in fact a finite sum since ϕ has compact support. Multiplying by h and rearranging this equation gives

$$hk\left[\sum_{n=1}^{\infty}\sum_{j=-\infty}^{\infty}\left(\frac{\phi(x_j, t_n) - \phi(x_j, t_{n-1})}{k}\right)U_j^n\right.\qquad(12.47)$$

$$+ \sum_{n=0}^{\infty} \sum_{j=-\infty}^{\infty} \left(\frac{\phi(x_{j+1}, t_n) - \phi(x_j, t_n)}{h} \right) F(U^n; j) \right] = -h \sum_{j=-\infty}^{\infty} \phi(x_j, 0) U_j^0.$$

This transformation using summation by parts is completely analogous to the derivation of (3.22) from (3.21).

Now let $l \to \infty$, so that k_l, $h_l \to 0$. At this point our simplified notation becomes difficult; the term U_j^n in (12.47) should be replaced by $U_l(x_j, t_n)$, for example, to explicitly show the dependece on l. It is reasonably straightforward, using the 1-norm convergence of U_l to u and the smoothness of ϕ, to show that the term on the top line of (12.47) converges to $\int_0^{\infty} \int_{-\infty}^{\infty} \phi_t(x, t) u(x, t) \, dx$ as $l \to \infty$. If we take initial data $U_j^0 = \bar{u}_j^0$, for example, then the right hand side converges to $- \int_{-\infty}^{\infty} \phi(x, 0) u(x, 0) \, dx$ as well.

The remaining term in (12.47), involving $F(U^n; j)$, is more subtle and requires the additional assumptions on F and U that we have imposed. The value of the numerical flux function F appearing here, which depends on some $p + q + 1$ values of U_l, can be written more properly as

$$F(U_l(x_j - ph, t_n), \ldots, U_l(x_j + qh, t_n)). \tag{12.48}$$

Since F is consistent with f, we have that

$$|F(U_l(x_j - ph, t_n), \ldots, U_l(x_j + qh, t_n)) - f(U_l(x_j, t_n))|$$
$$\leq K \max_{-p \leq i \leq q} |U_l(x_j + ih, t_n) - U_l(x_j, t_n)|$$

where K is the Lipschitz constant. Moreover, since $U_l(\cdot, t)$ has bounded total variation, uniformly in l, it must be that

$$\max_{-p \leq i \leq q} |U_l(x + ih, t) - U_l(x, t)| \to 0 \quad \text{as } l \to \infty$$

for almost all values of x. Consequently, we have that the numerical flux function (12.48) can be approximated by $f(U_l(x_j, t_n))$ with errors that vanish uniformly almost everywhere. This is the critical step in the proof, and together with some additional standard estimates gives the convergence of (12.47) to the weak form (12.43). Since this is true for any test function $\phi \in C_0^1$, we have proved that u is in fact a weak solution.

12.5 The entropy condition

This theorem does not guarantee that weak solutions obtained in this manner satisfy the entropy condition, and there are many examples of conservative numerical methods that converge to weak solutions violating the entropy condition.

EXAMPLE 12.3. Consider Burgers' equation with data

$$u_0(x) = \begin{cases} -1 & x < 0 \\ +1 & x > 0 \end{cases} \tag{12.49}$$

which we might naturally discretize by setting

$$U_j^0 = \begin{cases} -1 & j \leq 0 \\ +1 & j > 0 \end{cases} \tag{12.50}$$

The entropy satisfying weak solution consists of a rarefaction wave, but the stationary discontinuity $u(x,t) = u_0(x)$ (for all x and t) is also a weak solution. The Rankine-Hugoniot condition with $s = 0$ is satisfied since $f(-1) = f(1)$ for Burgers' equation. There are very natural conservative methods that converge to this latter solution rather than to the physically correct rarefaction wave.

One example is a natural generalization of the upwind methods (12.13) and (12.14) given by

$$F(v,w) = \begin{cases} f(v) & \text{if} \quad (f(v) - f(w))/(v - w) \geq 0 \\ f(w) & \text{if} \quad (f(v) - f(w))/(v - w) < 0. \end{cases} \tag{12.51}$$

This attempts to use the appropriate "upwind" direction even for problems where this direction changes from point to point. In many cases this works adequately. However, for the problem we consider here we will obtain $U_j^{n+1} = U_j^n$ for all j and n since $f(-1) = f(1)$ and so all the flux differences cancel out. Consequently $U_k(x,t) = U_k(x,0)$ for all t, and we converge to $u(x,t) = u_0(x)$ as the grid is refined.

Note the sensitivity of this numerical solution to our choice of initial data. If we instead take a different discretization of (12.49), say

$$U_j^0 = \begin{cases} -1 & j < 0 \\ 0 & j = 0 \\ +1 & j > 0 \end{cases} \tag{12.52}$$

then it turns out that the upwind method (12.51) gives the proper rarefaction wave solution.

The physically correct solution to this problem is often called a **transonic rarefaction** because of the fact that the wave speed passes through zero within the rarefaction wave. (In the Euler equations, this only happens when an eigenvalue $v \pm c$ passes through zero, where v is the fluid velocity and c is the sound speed. This means that the flow is subsonic to one side and supersonic to the other.) It is in precisely this situation where entropy violating shocks are most frequently computed.

For some numerical methods, it is possible to show that this can never happen, and that any weak solution obtained by refining the grid must in fact satisfy the entropy condition. Of course this supposes that we have a suitable entropy condition for the system to begin with, and the most convenient form is typically the entropy inequality. Recall that this requires a scalar entropy function $\eta(u)$ and entropy flux $\psi(u)$ for which

$$\frac{\partial}{\partial t}\eta\left(u(x,t)\right) + \frac{\partial}{\partial x}\psi\left(u(x,t)\right) \leq 0 \tag{12.53}$$

in the weak sense (version IV of the entropy condition). This is equivalent to the statement that

$$\int_0^\infty \int_{-\infty}^\infty \phi_t \eta(u) + \phi_x \psi(u) \, dx \, dt \le - \int_{-\infty}^\infty \phi(x, 0) \eta(u(x, 0)) \, dx \qquad (12.54)$$

for all $\phi \in C_0^1$ with $\phi(x, t) \ge 0$ for all x, t.

In order to show that the weak solution $u(x, t)$ obtained as the limit of $U_l(x, t)$ satisfies this inequality, it suffices to show that a discrete entropy inequality holds, of the form

$$\eta(U_j^{n+1}) \le \eta(U_j^n) - \frac{k}{h} [\Psi(U^n; j) - \Psi(U^n; j - 1)]. \qquad (12.55)$$

Here Ψ is some numerical entropy flux function that must be consistent with ψ in the same manner that we require F to be consistent with f. If we can show that (12.55) holds for a suitable Ψ, then mimicking the proof of the Lax-Wendroff Theorem (i.e., multiplying (12.55) by $\phi(x_j, t_n)$, summing over j and n, and using summation by parts), we can show that the limiting weak solution $u(x, t)$ obtained as the grid is refined satisfies the entropy inequality (12.54).

In the next chapter we will study a version of the upwind method for which a discrete entropy inequality of this form can be easily proved.

EXERCISE 12.4. *Consider the upwind method with flux (12.51). Take $k/h = 1/2$ and apply it to Burgers' equation with initial data*

$$u_0(x) = \begin{cases} -1 & x < 1 \\ +1 & x > 1 \end{cases} \qquad (12.56)$$

discretized using $U_j^0 = \bar{u}_j^0$ (cell averages). Based on the behavior of this method as described above, justify the following statements:

1. *The sequence $U_l(x, t)$ for $k_l = 1/2l$ converges to the correct rarefaction wave solution as $l \to \infty$.*

2. *The sequence $U_l(x, t)$ for $k_l = 1/(2l + 1)$ converges to an entropy violating shock as $l \to \infty$.*

3. *The sequence $U_l(x, t)$ for $k_l = 1/l$ does not converge as $l \to \infty$.*

13 Godunov's Method

Recall that one-sided methods cannot be used for systems of equations with eigenvalues of mixed sign. For a linear system of equations we previously obtained a natural generalization of the upwind method by diagonalizing the system, yielding the method (10.60). For nonlinear systems the matrix of eigenvectors is not constant, and this same approach does not work directly. In this chapter we will study a generalization in which the local characteristic structure, now obtained by solving a Riemann problem rather than by diagonalizing the Jacobian matrix, is used to define a natural upwind method. This method was first proposed for gas dynamics calculations by Godunov[24].

Since solving a Riemann problem is a nontrivial task, we should first justify the need for upwind methods. An alternative is to use a centered method such as the Lax-Friedrichs method (12.15) which allows eigenvalues of either sign. In the scalar case, this turns out to be stable provided

$$\left| \frac{k}{h} f'(U_j^n) \right| \leq 1 \tag{13.1}$$

for all U_j^n. However, in Chapter 11 we have seen for the linear advection equation that Lax-Friedrichs is generally more dissipative than the upwind method, and gives less accurate solutions. This is not surprising since the theory of characteristics tells us that the solution at x_j depends only on the data to the left of x_j and the upwind method takes advantage of this knowledge. In fact, for the advection equation we know that

$$u(x_j, t_{n+1}) = u(x_j - ak, t_n) \tag{13.2}$$

and both upwind and Lax-Friedrichs can be viewed as interpolation formulas to approximate u at $x_j - ak$ from the values of u at the grid points. Upwind is better because it interpolates using values at the two nearest grid points x_{j-1} and x_j,

$$U_j^{n+1} = \frac{1}{h} \left[(h - ak)U_j^n + akU_{j-1}^n \right] \tag{13.3}$$

whereas Lax-Friedrichs interpolates using values at x_{j-1} and x_{j+1},

$$U_j^{n+1} = \frac{1}{2h} \left[(h - ak)U_{j+1}^n + (h + ak)U_{j-1}^n \right]. \tag{13.4}$$

By doing upwind differencing in the appropriate direction for each characteristic component in a system of equations we can hope to obtain a similar improvement and decrease the numerical dissipation.

Looking ahead, there are other reasons for introducing methods based on the solution of Riemann problems. Both Lax-Friedrichs and the upwind method are only first order accurate on smooth data, and even the less dissipative upwind method gives unacceptably smeared shock profiles. We ultimately want to correct these deficiencies by developing "high resolution" methods that are second order accurate in smooth regions and give much sharper discontinuities. However, our experience with the linear advection equation (e.g., in Figure 11.1) indicates that natural second order methods, even one-sided methods like Beam-Warming, give oscillatory solutions. We will be able to cure this only by using more information about the local behavior of the solution. The first order Godunov method introduced here forms a basis for many of the high resolution generalizations that will be studied later.

EXERCISE 13.1. *Show that the Lax-Wendroff method on $u_t + au_x = 0$ can be derived by approximating $u(x_j - ak, t_n)$ using quadratic interpolation based on the points U_{j-1}^n, U_j^n, U_{j+1}^n. Use this interpretation to explain why oscillations appear near a discontinuity with Lax-Wendroff but not with Upwind or Lax-Friedrichs. Similarly, Beam-Warming corresponds to quadratic interpolation based on U_{j-2}^n, U_{j-1}^n, U_j^n.*

13.1 The Courant-Isaacson-Rees method

Historically, one of the first attempts at upwinding for the equations of gas dynamics was made by Courant, Isaacson and Rees[14] in 1952. They proposed solving certain equations along the characteristics going back from the point (x_j, t_{n+1}). To evaluate the characteristic variables at time t_n, this method uses interpolation based on the two nearest grid values, which are (U_{j-1}^n, U_j^n) or (U_j^n, U_{j+1}^n) depending on whether the corresponding characteristic speed is positive or negative. Of course the exact path of the characteristic is not known, but is approximated by a straight line with slope $\lambda_p(U_j^n)$. This is schematically illustrated in Figure 13.1.

For the scalar advection equation this reduces to the upwind method. For a scalar nonlinear problem, in which u is constant along characteristics, it reduces to determining U_j^{n+1} by an approximation to U at the point $x_j - f'(U_j^n)k$ obtained by linear interpolation of the data U^n. For example, if $f'(U_j^n) > 0$ then we would interpolate between U_{j-1}^n and U_j^n:

$$U_j^{n+1} = \frac{1}{h}\left[(h - f'(U_j^n)k)U_j^n + f'(U_j^n)kU_{j-1}^n\right] \tag{13.5}$$

$$= U_j^n - \frac{k}{h}f'(U_j^n)\left[U_j^n - U_{j-1}^n\right].$$

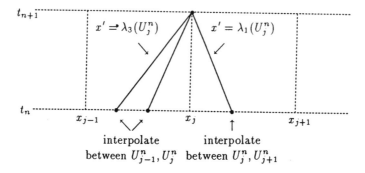

Figure 13.1. In the CIR method, characteristics are traced back from (x_j, t_{n+1}) and linear interpolation used at time t_n.

This is a natural upwind approximation to $u_t + f'(u)u_x = 0$, but this is not a good method for a problem involving shocks, since it is not in conservation form. (Recall our results with the method (12.4), which is precisely the CIR method (13.5) on Burgers' equation.)

13.2   Godunov's method

In 1959, Godunov[24] proposed a way to make use of the characteristic information within the framework of a conservative method. Rather than attempting to follow characteristics backwards in time, Godunov suggested solving Riemann problems forward in time. Solutions to Riemann problems are relatively easy to compute, give substantial information about the characteristic structure, and lead to conservative methods since they are themselves exact solutions of the conservation laws and hence conservative.

In Godunov's method, we use the numerical solution U^n to define a piecewise constant function $\tilde{u}^n(x, t_n)$ with the value U_j^n on the grid cell $x_{j-1/2} < x < x_{j+1/2}$. At time t_n this agrees with the piecewise constant function $U_k(x, t_n)$ that has already been introduced, but the function $\tilde{u}^n$, unlike U_k, will not be constant over $t_n \le t < t_{n+1}$. Instead, we use $\tilde{u}^n(x, t_n)$ as initial data for the conservation law, which we now solve exactly to obtain $\tilde{u}^n(x, t)$ for $t_n \le t \le t_{n+1}$. The equation can be solved exactly over a short time interval because the initial data $\tilde{u}^n(x, t_n)$ is piecewise constant, and hence defines a sequence of Riemann problems. The exact solution, up to the time when waves from neighboring

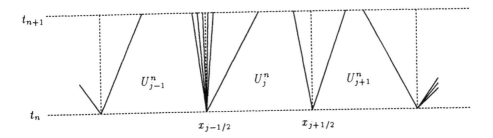

Figure 13.2. Solution of independent Riemann problems for the piecewise constant data $\tilde{u}^n(x, t_n)$ in the case $m = 2$.

Riemann problems begin to interact, is obtained by simply piecing together these Riemann solutions. This is illustrated in Figure 13.2.

After obtaining this solution over the interval $[t_n, t_{n+1}]$, we define the approximate solution U^{n+1} at time t_{n+1} by averaging this exact solution at time t_{n+1},

$$U_j^{n+1} = \frac{1}{h} \int_{x_{j-1/2}}^{x_{j+1/2}} \tilde{u}^n(x, t_{n+1}) \, dx. \tag{13.6}$$

These values are then used to define new piecewise constant data $\tilde{u}^{n+1}(x, t_{n+1})$ and the process repeats.

In practice this algorithm is considerably simplified by observing that the cell average (13.6) can be easily computed using the integral form of the conservation law. Since $\tilde{u}^n$ is assumed to be an exact weak solution, we know that

$$\int_{x_{j-1/2}}^{x_{j+1/2}} \tilde{u}^n(x, t_{n+1}) \, dx = \int_{x_{j-1/2}}^{x_{j+1/2}} \tilde{u}^n(x, t_n) \, dx + \int_{t_n}^{t_{n+1}} f(\tilde{u}^n(x_{j-1/2}, t)) \, dt$$
$$- \int_{t_n}^{t_{n+1}} f(\tilde{u}^n(x_{j+1/2}, t)) \, dt.$$

Dividing by h, using (13.6), and noting that $\tilde{u}^n(x, t_n) \equiv U_j^n$ over the cell $(x_{j-1/2}, x_{j+1/2})$, this equation reduces to

$$U_j^{n+1} = U_j^n - \frac{k}{h} \left[F(U_j^n, U_{j+1}^n) - F(U_{j-1}^n, U_j^n) \right] \tag{13.7}$$

where the numerical flux function F is given by

$$F(U_j^n, U_{j+1}^n) = \frac{1}{k} \int_{t_n}^{t_{n+1}} f\left(\tilde{u}^n(x_{j+1/2}, t)\right) dt. \tag{13.8}$$

This shows that Godunov's method can be written in conservation form. Moreover, note that the integral we need to compute in (13.8) is trivial because $\tilde{u}^n$ is *constant* at the point $x_{j+1/2}$ over the time interval (t_n, t_{n+1}). This follows from the fact that the solution of the Riemann problem at $x_{j+1/2}$ is a similarity solution, constant along each ray $(x - x_{j+1/2})/t = $ constant.

The constant value of $\tilde{u}^n$ along the line $x = x_{j+1/2}$ depends only on the data U_j^n and U_{j+1}^n for this Riemann problem. If we denote this value by $u^*(U_j^n, U_{j+1}^n)$, then the flux (13.8) reduces to

$$F(U_j^n, U_{j+1}^n) = f\left(u^*(U_j^n, U_{j+1}^n)\right) \tag{13.9}$$

and Godunov's method becomes

$$U_j^{n+1} = U_j^n - \frac{k}{h}\left[f\left(u^*(U_j^n, U_{j+1}^n)\right) - f\left(u^*(U_{j-1}^n, U_j^n)\right)\right]. \tag{13.10}$$

Note that the flux (13.9) is consistent with f since if $U_j^n = U_{j+1}^n \equiv \bar{u}$ then $u^*(U_j^n, U_{j+1}^n) = \bar{u}$ as well. Lipschitz continuity follows from smoothness of f.

For large t, of course, the solution may not remain constant at $x_{j+1/2}$ because of the effect of waves arising from neighboring Riemann problems. However, since the wave speeds are bounded by the eigenvalues of $f'(u)$ and the neighboring Riemann problems are distance h away, $\tilde{u}^n(x_{j+1/2}, t)$ will be constant over $[t_n, t_{n+1}]$ provided k is sufficiently small. We require that

$$\left|\frac{k}{h}\lambda_p(U_j^n)\right| \le 1 \tag{13.11}$$

for all eigenvalues λ_p at each U_j^n. The maximum of this quantity over the values of u arising in a particular problem is called the **Courant number**, a natural generalization of (10.56) from the linear problem.

Note that (13.11) allows the interaction of waves from neighboring Riemann problems during the time step, provided the interaction is entirely contained within a mesh cell. See Figure 13.3 for an example. In this case the solution $\tilde{u}^n(x, t)$ would be difficult or impossible to calculate. However, we never explicitly calculate the full solution, since all we require is the cell average (13.6). This is still easy to compute because $\tilde{u}^n$ remains constant on each cell boundary.

13.3   Linear systems

For a constant coefficient linear system $u_t + Au_x = 0$, solving the Riemann problem with left and right states U_j^n and U_{j+1}^n gives an intermediate value $u^*(U_j^n, U_{j+1}^n)$ that can be

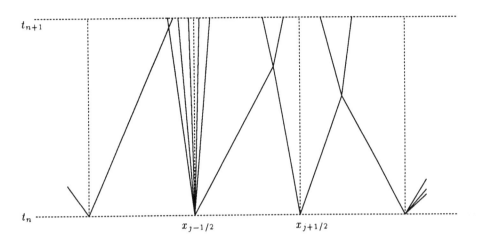

Figure 13.3. Exact solution $\tilde{u}^n$ used in determining U_j^{n+1} by Godunov's method. Note that $\tilde{u}^n$ is constant along the line segments $t_n < t < t_{n+1}$ at $x = x_{j-1/2}$ and $x = x_{j+1/2}$.

written in two different ways, as

$$u^*(U_j^n, U_{j+1}^n) = U_j^n + \sum_{\lambda_p < 0} \alpha_p r_p = U_{j+1}^n - \sum_{\lambda_p > 0} \alpha_p r_p. \tag{13.12}$$

Here r_p is the pth eigenvector of A and α_p is the coefficient of r_p in an eigenvector expansion of $U_{j+1}^n - U_j^n$, so the vector of α_p's is given by $\alpha = R^{-1}(U_{j+1}^n - U_j^n)$. Using the notation of (10.61), we can write the corresponding flux for Godunov's method as

$$
\begin{aligned}
F(U_j^n, U_{j+1}^n) &= Au^*(U_j^n, U_{j+1}^n) \\
&= AU_j^n + \sum_{\lambda_p < 0} \alpha_p \lambda_p r_p = AU_{j+1}^n - \sum_{\lambda_p > 0} \alpha_p \lambda_p r_p \\
&= AU_j^n + A^-(U_{j+1}^n - U_j^n) = AU_{j+1}^n - A^+(U_{j+1}^n - U_j^n).
\end{aligned}
\tag{13.13}
$$

If we choose the first of these equivalent expressions for $F(U_j^n, U_{j+1}^n)$ and take the second to define $F(U_{j-1}^n, U_j^n)$, i.e.,

$$F(U_{j-1}^n, U_j^n) = AU_j^n - A^+(U_j^n - U_{j-1}^n), \tag{13.14}$$

then Godunov's method for the linear system takes the form

$$
\begin{aligned}
U_j^{n+1} &= U_j^n - \frac{k}{h}[F(U_j^n, U_{j+1}^n) - F(U_{j-1}^n, U_j^n)] \\
&= U_j^n - \frac{k}{h}[A^-(U_{j+1}^n - U_j^n) + A^+(U_j^n - U_{j-1}^n)].
\end{aligned}
$$

This is simply the upwind method (10.60) for the linear system.

Notice that yet another expression for the flux $F(U_j^n, U_{j+1}^n)$ can be obtained by averaging the two expressions given in (13.13), obtaining

$$
\begin{aligned}
F(U_j^n, U_{j+1}^n) &= \frac{1}{2}A(U_j^n + U_{j+1}^n) + \frac{1}{2}(A^- - A^+)(U_{j+1}^n - U_j^n) \\
&= \frac{1}{2}A(U_j^n + U_{j+1}^n) - \frac{1}{2}|A|(U_{j+1}^n - U_j^n)
\end{aligned}
\tag{13.15}
$$

where

$$
|A| = A^+ - A^- = R|\Lambda|R^{-1} \qquad \text{with } |\Lambda| = \operatorname{diag}(|\lambda_1|, \ldots, |\lambda_m|).
\tag{13.16}
$$

If we use this form in Godunov's method we obtain an alternative formulation of the linear upwind method,

$$
U_j^{n+1} = U_j^n - \frac{k}{2h}A(U_{j+1}^n - U_{j-1}^n) + \frac{k}{2h}|A|(U_{j+1}^n - 2U_j^n + U_{j-1}^n).
\tag{13.17}
$$

The first two terms here correspond to the centered method (10.7) which is unconditionally unstable, but the last term is a dissipative term that stabilizes the method. Note that $(U_{j+1}^n - 2U_j^n + U_{j-1}^n) \approx h^2 u_{xx}(x_j, t_n)$ and that the diffusion matrix $|A|$ has nonnegative eigenvalues. This formulation of the upwind method for linear systems will prove useful later.

13.4   The entropy condition

The function $\tilde{u}^n(x, t)$ for $t_n \leq t \leq t_{n+1}$ is assumed to be a weak solution of the conservation law. In situations where this weak solution is not unique, there may be several choices for $\tilde{u}^n$. Different choices may give different values of $u^*(U_j^n, U_{j+1}^n)$ and hence different numerical solutions. The method is conservative and consistent regardless of what choice we make, but in cases where there is a unique weak solution that satisfies some physically motivated entropy condition, it makes sense to use the entropy-satisfying weak solution for $\tilde{u}^n$ in each time step. We might hope that by doing this the numerical solution will satisfy a discrete version of the entropy condition. This is in fact true, as we verify below. It then follows from the theory of Chapter 12 that any limiting function obtained by refining the grid must then be an entropy-satisfying weak solution. (If, on the other hand, we use Riemann solutions that do not satisfy the entropy condition in defining $u^*(U_j^n, U_{j+1}^n)$, then our numerical solution may converge to a weak solution that does not satisfy the entropy condition.)

Suppose that we have a convex entropy function $\eta(u)$ and entropy flux $\psi(u)$, and that every $\tilde{u}^n$ satisfies the entropy inequality (12.53). Then we wish to derive a discrete entropy inequality of the form (12.55) for Godunov's method. Since $\tilde{u}^n(x,t)$ for $t_n \leq t \leq t_{n+1}$ represents the *exact* entropy satisfying solution, we can integrate (12.53) over the rectangle $(x_{j-1/2}, x_{j+1/2}) \times (t_n, t_{n+1})$ to obtain

$$\frac{1}{h} \int_{x_{j-1/2}}^{x_{j+1/2}} \eta\left(\tilde{u}^n(x, t_{n+1})\right)\, dx \leq \frac{1}{h} \int_{x_{j-1/2}}^{x_{j+1/2}} \eta\left(\tilde{u}^n(x, t_n)\right)\, dx$$
$$-\frac{1}{h} \left[\int_{t_n}^{t_{n+1}} \psi\left(\tilde{u}^n(x_{j+1/2}, t)\right)\, dt - \int_{t_n}^{t_{n+1}} \psi\left(\tilde{u}^n(x_{j-1/2}, t)\right)\, dt\right].$$

This is almost what we need. Since $\tilde{u}^n$ is constant along three of the four sides of this rectangle, all integrals on the right hand side can be evaluated to give

$$\frac{1}{h} \int_{x_{j-1/2}}^{x_{j+1/2}} \eta\left(\tilde{u}^n(x, t_{n+1})\right)\, dx \;\leq\; \eta(U_j^n) \tag{13.18}$$
$$-\frac{k}{h}\left[\psi\left(u^*(U_j^n, U_{j+1}^n)\right) - \psi\left(u^*(U_{j-1}^n, U_j^n)\right)\right]$$

Again u^* represents the value propagating with velocity 0 in the solution of the Riemann problem. If we define the numerical entropy flux by

$$\Psi(U_j^n, U_{j+1}^n) = \psi\left(u^*(U_j^n, U_{j+1}^n)\right), \tag{13.19}$$

then Ψ is consistent with ψ, and the right hand side of (13.18) agrees with (12.55).

The left hand side of (13.19) is not equal to $\eta(U_j^{n+1})$, because $\tilde{u}^n$ is not constant in this interval. However, since the entropy function η is convex, we can use *Jensen's inequality* which says that the value of η evaluated at the average value of $\tilde{u}^n$ is less than or equal to the average value of $\eta(\tilde{u}^n)$, i.e.,

$$\eta\left(\frac{1}{h} \int_{x_{j-1/2}}^{x_{j+1/2}} \tilde{u}^n(x, t_{n+1})\, dx\right) \leq \frac{1}{h} \int_{x_{j-1/2}}^{x_{j+1/2}} \eta\left(\tilde{u}^n(x, t_{n+1})\right)\, dx. \tag{13.20}$$

The left hand side here is simply $\eta(U_j^{n+1})$ while the right hand side is bounded by (13.18). Combining (13.18), (13.19) and (13.20) thus gives the desired entropy inequality

$$\eta(U_j^{n+1}) \leq \eta(U_j^n) - \frac{k}{h}\left[\Psi\left(U_j^n, U_{j+1}^n\right) - \Psi\left(U_{j-1}^n, U_j^n\right)\right].$$

This shows that weak solutions obtained by Godunov's method satisfy the entropy condition, provided we use entropy-satisfying Riemann solutions.

13.5 Scalar conservation laws

If we apply Godunov's method to a scalar nonlinear equation, we obtain generalizations of the upwind methods defined by (12.13) and (12.14) for the case where the sign of $f'(u)$

varies. Recall that the Riemann problem with data u_l, u_r always has one weak solution consisting simply of the discontinuity propagating with speed $s = (f(u_r) - f(u_l))/(u_r - u_l)$. If we always use this Riemann solution (which may not satisfy the entropy condition), then the intermediate value u^* in the Riemann solution is given by

$$u^*(u_l, u_r) = \begin{cases} u_l & \text{if } s > 0, \\ u_r & \text{if } s < 0. \end{cases} \tag{13.21}$$

If $s = 0$ then this value is not well defined, but note that in this case $f(u_l) = f(u_r)$ and so the resulting flux, which is all that is required in Godunov's method, is the same whether we define $u^* = u_l$ or $u^* = u_r$.

From (13.21) we thus find the flux function

$$\begin{aligned} F(u_l, u_r) &= f(u^*(u_l, u_r)) \\ &= \begin{cases} f(u_l) & \text{if } (f(u_r) - f(u_l))/(u_r - u_l) \geq 0, \\ f(u_r) & \text{if } (f(u_r) - f(u_l))/(u_r - u_l) < 0 \end{cases} \end{aligned} \tag{13.22}$$

Notice that this is precisely the flux (12.51) which, as we have already seen, gives a method that may compute entropy-violating solutions.

To fix this, we should use the entropy-satisfying weak solution in implementing Godunov's method. This may consist of a rarefaction wave rather than a shock (or of some combination of shocks and rarefactions in the nonconvex case).

In the convex case, there are four cases that must be considered. The reader should verify the following results in each case:

1. $f'(u_l), f'(u_r) \geq 0 \implies u^* = u_l$,

2. $f'(u_l), f'(u_r) \leq 0 \implies u^* = u_r$,

3. $f'(u_l) \geq 0 \geq f'(u_r) \implies u^* = u_l$ if $[f]/[u] > 0$ or $u^* = u_r$ if $[f]/[u] < 0$,

4. $f'(u_l) < 0 < f'(u_r) \implies u^* = u_s$ (transonic rarefaction).

In each of the first three cases, the value u^* is either u_l or u_r and the flux is correctly given by (13.22). Note in particular that in Cases 1 and 2 it is irrelevant whether the solution is a shock or rarefaction, since the value of u^* is the same in either case. This shows that using Godunov's method with entropy-violating Riemann solutions does not necessarily lead to entropy-violating numerical solutions!

It is only in Case 4, the transonic rarefaction, that the value of F differs from (13.22). In this case u^* is neither u_l nor u_r, but is the intermediate value u_s with the property that

$$f'(u_s) = 0. \tag{13.23}$$

This is the value of u for which the characteristic speed is zero, and is called the sonic point.

We can modify (13.22) to include this possibility: if $f'(u_l) < 0 < f'(u_r)$ then we redefine

$$F(u_l, u_r) = f(u_s).$$

The resulting flux function can be written in a simplified form, however, as

$$F(u_l, u_r) = \begin{cases} \min_{u_l \leq u \leq u_r} f(u) & \text{if } u_l \leq u_r \\ \max_{u_r \leq u \leq u_l} f(u) & \text{if } u_l > u_r. \end{cases} \tag{13.24}$$

EXERCISE 13.2. *Verify this by considering all possible cases.*

More surprisingly, it turns out that (13.24) is valid more generally for any scalar conservation law, even nonconvex ones, and gives the correct Godunov flux corresponding to the weak solution satisfying Olienik's entropy condition (3.46).

EXERCISE 13.3. *Verify this claim, using the convex hull construction of the entropy-satisfying Riemann solution presented in Chapter 4.*

This also follows from a more general result due to Osher[58], who found a closed form expression for the entropy solution $u(x, t) \equiv w(x/t)$ of a general nonconvex scalar Riemann problem with data u_l and u_r. The solution $w(x/t) = w(\xi)$ satisfies the implicit relation

$$f(w(\xi)) - \xi w(\xi) = g(\xi) \equiv \begin{cases} \min_{u_l \leq u \leq u_r} [f(u) - \xi u] & \text{if } u_l \leq u_r \\ \max_{u_r \leq u \leq u_l} [f(u) - \xi u] & \text{if } u_l > u_r. \end{cases} \tag{13.25}$$

Setting $\xi = 0$ gives $f(w(0)) = f(u^*(u_l, u_r))$ which is precisely the numerical flux (13.24). Osher goes on to show that an explicit expression for $w(\xi)$ is obtained by differentiating the above expression, yielding

$$w(\xi) = -g'(\xi) \tag{13.26}$$

where $g(\xi)$ is the function defined by the right side of (13.25).

14 Approximate Riemann Solvers

Godunov's method, and higher order variations of the method to be discussed later, require the solution of Riemann problems at every cell boundary in each time step. Although in theory these Riemann problems can be solved, in practice doing so is expensive, and typically requires some iteration for nonlinear equations.

Note that most of the structure of the resulting Riemann solver is not used in Godunov's method. The exact solution is averaged over each grid cell, introducing large numerical errors. This suggests that it is not worthwhile calculating the Riemann solutions exactly and that we may be able to obtain equally good numerical results with an approximate Riemann solution obtained by some less expensive means.

There are two distinct ways we could think of generalizing Godunov's method. One is to start with the Godunov flux

$$F(u_l, u_r) = f(u^*(u_l, u_r)) \tag{14.1}$$

where $u^*(u_l, u_r)$ is the intermediate state $w(0)$ arising in the similarity solution $u(x, t) = w(x/t)$ of the Riemann problem, and replace the function $u^*(u_l, u_r)$ by some approximation $\hat{u}^*(u_l, u_r)$. This leads to the approximate Godunov method

$$U_j^{n+1} = U_j^n - \frac{k}{h} \left[f\left(\hat{u}^*(U_j^n, U_{j+1}^n) \right) - f\left(\hat{u}^*(U_{j-1}^n, U_j^n) \right) \right]. \tag{14.2}$$

Note that this method will be conservative and consistent for any choice of the function $\hat{u}^*$, provided it satisfies the natural condition that $\hat{u}^*(\bar{u}, \bar{u}) = \bar{u}$ with appropriate Lipschitz continuity.

We might define the function $\hat{u}^*$ by first defining an approximate Riemann solution $\hat{u}(x, t) = \hat{w}(x/t)$ (see below) and then setting $\hat{u}^*(u_l, u_r) = \hat{w}(0)$.

A second approach to generalizing Godunov's method is to go back to the original description of the method, where U_j^{n+1} is taken to be the cell average of $\tilde{u}^n(x, t_{n+1})$, and now replace $\tilde{u}^n(x, t)$ by an approximate solution $\hat{u}^n(x, t)$. This can be defined by simply piecing together approximate Riemann solutions at each cell interface, just as $\tilde{u}^n$ was defined for the exact Riemann solutions. This requires that we define an approximate

Riemann solution $\hat{w}(x/t)$ with finite propagation speeds, so that

$$\hat{w}(\xi) = \begin{cases} u_l & \text{for } \xi < a_{\min} \\ u_r & \text{for } \xi > a_{\max} \end{cases} \tag{14.3}$$

where $a_{\min}$ and $a_{\max}$ are the minimum and maximum propagation speeds (which typically correspond closely to the range of eigenvalues of $f'(u_l)$ and $f'(u_r)$). We need

$$\left|\frac{ak}{h}\right| < \frac{1}{2} \quad \text{for all } a \text{ between } a_{\min} \text{ and } a_{\max} \tag{14.4}$$

in order for the construction of $\hat{u}^n(x,t)$ described above to work, but once we determine the corresponding flux function this can be relaxed to require only $|ak/h| < 1$ as in Godunov's method with the exact Riemann solution.

Once the approximate solution $\hat{u}^n$ is defined in the strip $t_n \le t \le t_{n+1}$, we can take U_j^{n+1} to be the cell average at time t_{n+1}:

$$U_j^{n+1} = \frac{1}{h} \int_{x_{j-1/2}}^{x_{j+1/2}} \hat{u}^n(x, t_{n+1}) \, dx. \tag{14.5}$$

Note that this will **not** in general be the same method as (14.2) with $\hat{u}^* = \hat{w}(0)$. If $\hat{u}^n$ is not an exact solution, we can no longer integrate the conservation laws over the rectangle $(x_{j-1/2}, x_{j+1/2}) \times (t_n, t_{n+1})$ to obtain (14.2) from (14.5).

The second approach is the one more commonly used, but requires some care in defining the approximate Riemann solution, since for arbitrary choices of $\hat{w}$ there is no guarantee that (14.5) is even conservative. Harten, Lax and van Leer[33] present some of the general theory of such approximate Riemann solvers.

14.1 General theory

To be conservative, the approximate Riemann solution $\hat{u}(x,t) = \hat{w}(x/t)$ must have the following property that for M sufficiently large,

$$\int_{-M}^{M} \hat{w}(\xi) \, d\xi = M(u_l + u_r) + f(u_l) - f(u_r). \tag{14.6}$$

Note that the exact Riemann solution $w(x/t)$ has this property, as seen from the integral form of the conservation law over $[-M, M] \times [0, 1]$.

If $\hat{w}$ satisfies (14.6) then we can write the resulting method in the standard conservation form. In order to determine the numerical flux $F(u_l, u_r)$, consider Figure 14.1. Since by consistency the flux should reduce to $f(u)$ wherever u is constant, we can integrate $\hat{w}$ from 0 to M, for M sufficiently large, and expect to get

$$\int_{0}^{M} \hat{w}(\xi) \, d\xi = M u_r + F(u_l, u_r) - f(u_r). \tag{14.7}$$

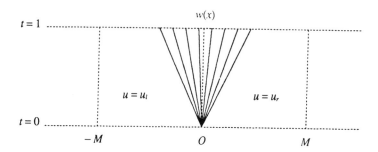

Figure 14.1. Integrating the approximate Riemann solution $\hat{w}$ gives the numerical flux.

Similarly, integrating from $-M$ to 0 gives

$$\int_{-M}^{0} \hat{w}(\xi)\, d\xi = M u_l + f(u_l) - F(u_l, u_r). \tag{14.8}$$

The condition (14.6) insures that these two requirements on F can be satisfied simultaneously, and solving for F gives two different expressions,

$$F(u_l, u_r) = f(u_r) - M u_r + \int_{0}^{M} \hat{w}(\xi)\, d\xi, \tag{14.9}$$

or, equivalently,

$$F(u_l, u_r) = f(u_l) + M u_l - \int_{-M}^{0} \hat{w}(\xi)\, d\xi. \tag{14.10}$$

14.1.1 The entropy condition

Harten and Lax[32] (see also [33]) point out that the Godunov-type method (14.5) is also consistent with the entropy condition provided the approximate Riemann solution satisfies another condition analogous to (14.6):

$$\int_{-M}^{M} \eta(\hat{w}(\xi))\, d\xi \leq M(\eta(u_l) + \eta(u_r)) + (\psi(u_l) - \psi(u_r)). \tag{14.11}$$

From this it can be shown that the discrete entropy condition (12.55) is satisfied with the numerical entropy flux

$$\Psi(u_l, u_r) = \psi(u_l) + M \eta(u_l) - \int_{-M}^{0} \eta(\hat{w}(\xi))\, d\xi. \tag{14.12}$$

Harten, Lax and van Leer present some approximate Riemann solvers that satisfy this condition.

14.1.2 Modified conservation laws

One natural way to obtain $\hat{w}(x/t)$ is to compute the *exact* Riemann solution to some modified conservation law $\hat{u}_t + \hat{f}(\hat{u})_x = 0$, with a flux function $\hat{f}(u)$ that is presumably easier to work with than the original flux $f(u)$. By using the integral form of this conservation law over $[-M, M] \times [0, 1]$, we see that the condition (14.6) will be satisfied provided that

$$\hat{f}(u_r) - \hat{f}(u_l) = f(u_r) - f(u_l). \tag{14.13}$$

The resulting numerical flux function is then given by

$$F(u_l, u_r) = \hat{f}(\hat{w}(0)) + f(u_r) - \hat{f}(u_r). \tag{14.14}$$

EXERCISE 14.1. *Verify (14.14).*

14.2 Roe's approximate Riemann solver

One of the most popular Riemann solvers currently in use is due to Roe[64]. The idea is to determine $\hat{u}(x, t)$ by solving a constant coefficient linear system of conservation laws instead of the original nonlinear system, *i.e.*, we solve a modified conservation law as described above with flux $\hat{f}(u) = \hat{A}u$. Of course the coefficient matrix used to define this linear system must depend on the u_l and u_r in order for (14.13) to be satisfied, so we will write the linear system for $\hat{u}$ as

$$\hat{u}_t + \hat{A}(u_l, u_r)\hat{u}_x = 0. \tag{14.15}$$

This linear Riemann problem is relatively easy to solve (see Section 6.5). If $\hat{A}$ has eigenvalues $\hat{\lambda}_i$ and eigenvectors $\hat{r}_i$, and if we decompose

$$u_r - u_l = \sum_p \alpha_p \hat{r}_p, \tag{14.16}$$

then

$$\hat{w}(\xi) = u_l + \sum_{\hat{\lambda}_p < \xi} \alpha_p \hat{r}_p \tag{14.17}$$

where the sum is over all p for which $\hat{\lambda}_p < \xi$. Equivalently,

$$\hat{w}(\xi) = u_r - \sum_{\hat{\lambda}_p > \xi} \alpha_p \hat{r}_p \tag{14.18}$$

We still have the problem of determining $\hat{A}(u_l, u_r)$ in a reasonable way. Roe suggested that the following conditions should be imposed on $\hat{A}$:

$$\begin{aligned}
&i) \quad \hat{A}(u_l, u_r)(u_r - u_l) = f(u_r) - f(u_l) \\
&ii) \quad \hat{A}(u_l, u_r) \text{ is diagonalizable with real eigenvalues} \\
&iii) \quad \hat{A}(u_l, u_r) \to f'(\bar{u}) \text{ smoothly as } u_l, u_r \to \bar{u}.
\end{aligned} \tag{14.19}$$

Condition (14.19*i*) has two effects. First, it is required by (14.13) and guarantees that the condition (14.6) is satisfied. Another effect is that, in the special case where u_l and u_r are connected by a single shock wave or contact discontinuity, the approximate Riemann solution agrees with the exact Riemann solution. This follows from the fact that the Rankine-Hugoniot condition is satisfied for u_l and u_r in this case, so

$$f(u_r) - f(u_l) = s(u_r - u_l)$$

for some s (the speed of the shock or contact). Combined with (14.19*i*), this shows that $u_r - u_l$ must, in this situation, be an eigenvector of $\hat{A}$ with eigenvalue s and so the approximate solution $\hat{u}(x, t)$ also consists of this single jump $u_r - u_l$ propagating with speed s.

Condition (14.19*ii*) is clearly required in order that the problem $\hat{u}_t + \hat{A}\hat{u}_x = 0$ is hyperbolic and solvable.

Condition (14.19*iii*) guarantees that the method behaves reasonably on smooth solutions, since if $\|U_j - U_{j-1}\| = O(h)$ then the linearized equation $u_t + f'(U_j)u_x = 0$ is approximately valid. It is natural to require that the linear system (14.15) agree with the linearization in this case. Since (14.19*i*) guarantees that the method behaves reasonably on an isolated discontinuity, it is only when a Riemann problem has a solution with more than one strong shock or contact discontinuity that the approximate Riemann solution will differ significantly from the true Riemann solution. In practice this happens infrequently — near the point where two shocks collide, for example.

One way to guarantee that both conditions (14.19*ii*) and (14.19*iii*) are satisfied is to take

$$\hat{A}(u_l, u_r) = f'(u_{\text{ave}}) \tag{14.20}$$

for some average value of u, e.g., $u_{\text{ave}} = \frac{1}{2}(u_l + u_r)$. Unfortunately, this simple choice of u_{ave} will not give an $\hat{A}$ that satisfies (14.19*i*) in general. Harten and Lax[32] show (see also Theorem 2.1 in [33]) that for a general system with an entropy function, a more complicated averaging of the Jacobian matrix in state space can be used. This shows that such linearizations exist, but is too complicated to use in practice.

Fortunately, for special systems of equations it is possible to derive suitable $\hat{A}$ matrices that are very efficient to use relative to the exact Riemann solution. Roe[64] showed how to do this for the Euler equations. Later in this chapter we will work through the analogous derivation for the isothermal equations.

14.2.1 The numerical flux function for Roe's solver

For Roe's approximate Riemann solver, the function $\hat{w}(x/t)$ is in fact the exact solution to a Riemann problem for the conservation law (14.15) with flux $\hat{f}(u) = \hat{A}u$. It follows from (14.14) that

$$
\begin{aligned}
F(u_l, u_r) &= \hat{A}\hat{w}(0) + f(u_r) - \hat{A}u_r \\
&= f(u_r) - \hat{A}\sum_{\hat{\lambda}_p > 0} \alpha_p \hat{r}_p \\
&= f(u_r) - \sum_{p=1}^{m} \hat{\lambda}_p^+ \alpha_p \hat{r}_p
\end{aligned}
\tag{14.21}
$$

where $\hat{\lambda}_p^+ = \max(\hat{\lambda}_p, 0)$. Alternatively, using (14.13) we obtain an equivalent expression,

$$
F(u_l, u_r) = f(u_l) + \sum_{p=1}^{m} \hat{\lambda}_p^- \alpha_p \hat{r}_p
\tag{14.22}
$$

where $\hat{\lambda}_p^- = \min(\hat{\lambda}_p, 0)$.

Note that if we average the two expressions (14.21) and (14.22), we obtain a third form,

$$
\begin{aligned}
F(u_l, u_r) &= \frac{1}{2}(f(u_l) + f(u_r)) - \sum_p |\hat{\lambda}_p| \alpha_p \hat{r}_p, \\
&= \frac{1}{2}(f(u_l) + f(u_r)) - |\hat{A}|(u_r - u_l),
\end{aligned}
\tag{14.23}
$$

where the absolute value of the matrix is defined as in (13.16). This form is reminiscent of the Godunov flux (13.15) for a linear system.

EXAMPLE 14.1.   For a scalar conservation law the condition (14.19i) determines $\hat{a} = \hat{A}(u_l, u_r)$ uniquely as

$$
\hat{a} = \frac{f(u_r) - f(u_l)}{u_r - u_l}.
\tag{14.24}
$$

The linearized problem is the scalar advection equation $\hat{u}_t + \hat{a}\hat{u}_x = 0$ and the approximate Riemann solution consists of the jump $u_r - u_l$ propagating with speed $\hat{a}$. But note that $\hat{a}$ in (14.24) is the Rankine-Hugoniot shock speed for the nonlinear problem, and so the "approximate" Riemann solution is in fact an exact weak solution (though one which may violate the entropy condition). Roe's method in the scalar case thus reduces to the method already discussed with flux (13.22), which we can rewrite using (14.22), for example, as

$$
F(u_l, u_r) = f(u_l) + \hat{a}^-(u_r - u_l).
\tag{14.25}
$$

14.2.2   A sonic entropy fix

One disadvantage of Roe's linearization is that the resulting approximate Riemann solution consists of only discontinuities, with no rarefaction waves. This can lead to a violation

of the entropy condition, as has been observed previously for the scalar method (13.22)
with appropriate initial data.

Recall that in the scalar case, the use of an entropy-violating Riemann solution leads
to difficulties only in the case of a sonic rarefaction wave, in which $f'(u_l) < 0 < f'(u_r)$.
This is also typically true when we use Roe's approximate Riemann solution for a system
of conservation laws. It is only for sonic rarefactions, those for which $\lambda_p < 0$ to the left
of the wave while $\lambda_p > 0$ to the right of the wave, that entropy violation is a problem.

In the case of a sonic rarefaction wave, it is necessary to modify the approximate
Riemann solver in order to obtain entropy satisfying solutions. There are various ways to
do this. One approach, discussed by Harten and Hyman[30], is outlined here.

For the wave in the pth family, traveling at speed $\hat{\lambda}_p$ according to the approximate
Riemann solver, compute the states to either side of this wave in $\hat{u}$:

$$u_{pl} = u_l + \sum_{i=1}^{p-1} \alpha_i \hat{r}_i, \qquad u_{pr} = u_{pl} + \alpha_p \hat{r}_p. \tag{14.26}$$

Also compute the true characteristic speed in the pth family for each of these states, say
$\lambda_{pl} = \lambda_p(u_{pl})$ and $\lambda_{pr} = \lambda_p(u_{pr})$. If $\lambda_{pl} > \lambda_{pr}$ then the characteristics are going into this
discontinuity. If $\lambda_{pl} < \lambda_{pr}$ then it should perhaps be a rarefaction wave, but it is only in
the transonic case where $\lambda_{pl} < 0 < \lambda_{pr}$ that we need to modify our approximate Riemann
solution.

Suppose that there appears to be a sonic rarefaction in the qth family for some q,
i.e., $\lambda_{ql} < 0 < \lambda_{qr}$. Then we replace the single jump $\hat{r}_q$ propagating at speed $\hat{\lambda}_q$ by two
jumps propagating at speeds λ_{ql} and λ_{qr}, with a new state u_{qm} in between. Conservation
requires that

$$(\lambda_{qr} - \lambda_{ql})u_{qm} = (\hat{\lambda}_q - \lambda_{ql})u_{ql} + (\lambda_{qr} - \hat{\lambda}_q)u_{qr}$$

so that the integral of $\hat{u}$ is unaffected by this modification. This determines u_{qm}:

$$u_{qm} = \frac{(\hat{\lambda}_q - \lambda_{ql})u_{ql} + (\lambda_{qr} - \hat{\lambda}_q)u_{qr}}{\lambda_{qr} - \lambda_{ql}}. \tag{14.27}$$

From this we can easily compute that $u_{qm} - u_{ql}$ and $u_{qr} - u_{qm}$ are both scalar multiples
of the eigenvector $\hat{r}_q$,

$$u_{qm} - u_{ql} = \left(\frac{\lambda_{qr} - \hat{\lambda}_q}{\lambda_{qr} - \lambda_{ql}}\right)\alpha_q \hat{r}_q, \tag{14.28}$$

$$u_{qr} - u_{qm} = \left(\frac{\hat{\lambda}_q - \lambda_{ql}}{\lambda_{qr} - \lambda_{ql}}\right)\alpha_q \hat{r}_q, \tag{14.29}$$

and of course the sum of these recovers $u_{qr} - u_{ql} = \alpha_q \hat{r}_q$. We can easily compute the
resulting flux function $F(u_l, u_r)$ when the entropy fix is applied to the qth family. If we

define

$$\hat{\lambda}_{ql} = \lambda_{ql} \left(\frac{\lambda_{qr} - \hat{\lambda}_q}{\lambda_{qr} - \lambda_{ql}} \right) \tag{14.30}$$

and

$$\hat{\lambda}_{qr} = \lambda_{qr} \left(\frac{\hat{\lambda}_q - \lambda_{ql}}{\lambda_{qr} - \lambda_{ql}} \right) \tag{14.31}$$

then

$$F(u_l, u_r) = f(u_l) + \sum_{p \neq q} \hat{\lambda}_p^- \alpha_p \hat{r}_p + \hat{\lambda}_{ql} \alpha_q \hat{r}_q \tag{14.32}$$

$$= f(u_r) - \sum_{p \neq q} \hat{\lambda}_p^+ \alpha_p \hat{r}_p - \hat{\lambda}_{qr} \alpha_q \hat{r}_q \tag{14.33}$$

This can be written more concisely if we make the following definitions for all $p = 1, 2, \ldots, m$:

$$\hat{\lambda}_{pl} = \lambda_{pl}^- \left(\frac{\lambda_{pr}^+ - \hat{\lambda}_p}{\lambda_{pr}^+ - \lambda_{pl}^-} \right) \tag{14.34}$$

and

$$\hat{\lambda}_{pr} = \lambda_{pr}^+ \left(\frac{\hat{\lambda}_p - \lambda_{pl}^-}{\lambda_{pr}^+ - \lambda_{pl}^-} \right). \tag{14.35}$$

In the sonic rarefaction case $\lambda_{pl}^- = \lambda_{pl}$ and $\lambda_{pr}^+ = \lambda_{pr}$ so these reduce to the formulas (14.30) and (14.31). In all other cases it is easy to verify that $\hat{\lambda}_{pl}$ reduces to $\hat{\lambda}_p^-$ while $\hat{\lambda}_{pr}$ reduces to $\hat{\lambda}_p^+$, so that we can write the flux function in the more general case as

$$F(u_l, u_r) = f(u_l) + \sum_p \hat{\lambda}_{pl} \alpha_p \hat{r}_p \tag{14.36}$$

$$= f(u_r) - \sum_p \hat{\lambda}_{pr} \alpha_p \hat{r}_p \tag{14.37}$$

with sonic rarefactions handled properly automatically.

EXERCISE 14.2. *Verify these claims.*

14.2.3 The scalar case

It is illuminating to consider the effect of this entropy fix in the scalar case. In this case there is only one characteristic family with eigenvalue $f'(u)$, and so we simplify notation by identifying $\lambda_{ql} = f'(u_l)$ and $\lambda_{qr} = f'(u_r)$. Moreover the Roe linearization gives the jump $(u_r - u_l)$ propagating with speed $\hat{\lambda}_q = \hat{a} = (f(u_r) - f(u_l))/(u_r - u_l)$.

In the sonic rarefaction case, $f'(u_l) < 0 < f'(u_r)$ and the entropy fix discussed above replaces the single jump propagating at speed $\hat{a}$ by two jumps propagating at speeds $f'(u_l)$

and $f'(u_r)$, separated by the state

$$u_m = \frac{(\hat{a} - f'(u_l))u_l + (f'(u_r) - \hat{a})u_r}{f'(u_r) - f'(u_l)}$$

$$= u_l + \left(\frac{f'(u_r) - \hat{a}}{f'(u_r) - f'(u_l)}\right)(u_r - u_l) \tag{14.38}$$

using (14.27). The approximate Riemann solution

$$\hat{w}(x/t) = \begin{cases} u_l & x/t < \hat{a} \\ u_r & x/t > \hat{a} \end{cases} \tag{14.39}$$

has been replaced by a modified approximate Riemann solution

$$\hat{w}(x/t) = \begin{cases} u_l & x/t < f'(u_l) \\ u_m & f'(u_l) < x/t < f'(u_r) \\ u_r & x/t > f'(u_r). \end{cases} \tag{14.40}$$

It is interesting to note that we can again interpret $\hat{w}(x/t)$ as the *exact* solution to a modified conservation law, now slightly more complicated than the advection equation $u_t + \hat{a}u_x = 0$ defining (14.39). The function $\hat{w}(x/t)$ in (14.40) is the exact Riemann solution to the problem $u_t + \hat{f}(u)_x = 0$, where

$$\hat{f}(u) = \begin{cases} f(u_l) + (u - u_l)f'(u_l) & u \le u_m \\ f(u_r) + (u - u_r)f'(u_r) & u \ge u_m. \end{cases} \tag{14.41}$$

This piecewise linear function is shown in Figure 14.2. It is easy to verify that the point u_m where the two tangent lines meet is the same point u_m defined by (14.38).

The flux function $\hat{f}(u)$ is not smooth at u_m, and so our standard theory does not apply directly to determine the solution to this Riemann problem. However, we can approximate $\hat{f}(u)$ arbitrarily well by smooth functions if we round off the corner at u_m, and in this way justify the claim that $\hat{w}$ in (14.40) solves this Riemann problem.

EXERCISE 14.3. *Determine the structure of the Riemann solution for a smoothed version of $\hat{f}(u)$ and verify that it consists of two contact discontinuities separated by a rarefaction wave, and approaches (14.40) in the limit of zero smoothing.*

The resulting numerical flux $F(u_l, u_r)$ can be evaluated from the general expression (14.32), which in the scalar case reduces to

$$F(u_l, u_r) = f(u_l) + f'(u_l)\left(\frac{f'(u_r) - \hat{a}}{f'(u_r) - f'(u_l)}\right)(u_r - u_l). \tag{14.42}$$

Alternatively, applying formula (14.14) in this case gives

$$F(u_l, u_r) = \hat{f}(u_m), \tag{14.43}$$

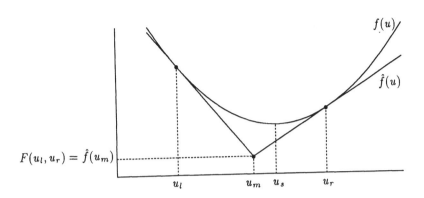

Figure 14.2. The piecewise linear function $\hat{f}(u)$ used to define the approximate Riemann solution in the sonic rarefaction case.

since $\hat{f}(u_r) = f(u_r)$. Using (14.41) we see that this agrees with (14.42).

This gives a nice geometric interpretation to the numerical flux in the scalar case when we apply the entropy fix. Moreover, we see from this that

$$F(u_l, u_r) < f(u_s) \tag{14.44}$$

where u_s is the sonic point at which $f'(u_s) = 0$ and f is a minimum. Recall that the Godunov flux using the correct rarefaction wave solution of the original conservation law is given by

$$F_G(u_l, u_r) = f(u_s) \tag{14.45}$$

in this case. Hence the flux we use in Roe's method with the entropy fix is always less than the Godunov flux. Since the flux for Roe's method always agrees with the Godunov flux in every case except the sonic rarefaction, we see that

$$
\begin{aligned}
F(u_l, u_r) &\leq F_G(u_l, u_r) \qquad \forall\, u_l \leq u_r, \\
F(u_l, u_r) &\geq F_G(u_l, u_r) \qquad \forall\, u_l \geq u_r,
\end{aligned}
\tag{14.46}
$$

where F_G denotes the Godunov flux. The second inequality of (14.46) is in fact satisfied as an equality for all $u_l \geq u_r$. It is written in the above form since it turns out that any numerical method for a scalar conservation law with a flux satisfying (14.46) automatically satisfies the entropy condition and can only converge to entropy-satisfying solutions. This shows that, in the scalar case at least, the entropy fix presented here does indeed cure the entropy problem.

Numerical methods satisfying (14.46) are called **E-schemes** (E for entropy), a notion introduced by Osher[58], who also discusses a related entropy fix for Roe's method and some alternative approximate Riemann solvers. See also Tadmor[85] for a discussion of these issues.

14.2.4   A Roe matrix for isothermal flow

As an example, we derive the Roe matrix for the isothermal equations of gas dynamics. Roe[64] presents the analogous formulas for the full Euler equations.

In the isothermal equations,

$$u = \begin{bmatrix} \rho \\ m \end{bmatrix}, \quad f(u) = \begin{bmatrix} m \\ m^2/\rho + a^2\rho \end{bmatrix}, \tag{14.47}$$

and

$$f'(u) = \begin{bmatrix} 0 & 1 \\ a^2 - m^2/\rho^2 & 2m/\rho \end{bmatrix} = \begin{bmatrix} 0 & 1 \\ a^2 - v^2 & 2v \end{bmatrix} \tag{14.48}$$

with $m = \rho v$. To derive an averaged Jacobian satisfying (14.19), introduce the new variables

$$z = \rho^{-1/2} u, \quad i.e. \quad \begin{bmatrix} z_1 \\ z_2 \end{bmatrix} = \begin{bmatrix} \rho^{1/2} \\ m/\rho^{1/2} \end{bmatrix} = \begin{bmatrix} \rho^{1/2} \\ \rho^{1/2} v \end{bmatrix}. \tag{14.49}$$

Then

$$u = z_1 z = \begin{bmatrix} z_1^2 \\ z_1 z_2 \end{bmatrix}, \quad f(u) = \begin{bmatrix} z_1 z_2 \\ a^2 z_1^2 + z_2^2 \end{bmatrix}. \tag{14.50}$$

The motivation for this change of variables is that both $(u_l - u_r)$ and $(f(u_l) - f(u_r))$ can be expressed in the form of some matrix times $(z_l - z_r)$. The averaging is done in terms of the variables z:

$$\bar{z} = \frac{1}{2}(z_l + z_r) = \begin{bmatrix} \bar{z}_1 \\ \bar{z}_2 \end{bmatrix} = \frac{1}{2} \begin{bmatrix} \rho_l^{1/2} + \rho_r^{1/2} \\ m_l/\rho_l^{1/2} + m_r/\rho_r^{1/2} \end{bmatrix}. \tag{14.51}$$

Then it is easy to check that

$$u_l - u_r = \begin{bmatrix} 2\bar{z}_1 & 0 \\ \bar{z}_2 & \bar{z}_1 \end{bmatrix} (z_l - z_r) \equiv \hat{B}(z_l - z_r),$$

$$f(u_l) - f(u_r) = \begin{bmatrix} \bar{z}_2 & \bar{z}_1 \\ 2a^2\bar{z}_1 & 2\bar{z}_2 \end{bmatrix} (z_l - z_r) \equiv \hat{C}(z_l - z_r).$$

Using [·] to denote the jump in a quantity, we have $[u] = \hat{B}[z]$ and $[f] = \hat{C}[z]$, which can be combined to give

$$[f] = \hat{C}\hat{B}^{-1}[u]. \tag{14.52}$$

Condition (14.19i) is thus satisfied if we take

$$\hat{A}(u_l, u_r) = \hat{C}\hat{B}^{-1} = \begin{bmatrix} 0 & 1 \\ a^2 - \bar{z}_2^2/\bar{z}_1^2 & 2\bar{z}_2/\bar{z}_1 \end{bmatrix} = \begin{bmatrix} 0 & 1 \\ a^2 - \bar{v}^2 & 2\bar{v} \end{bmatrix} \tag{14.53}$$

where we have defined the average velocity $\bar{v}$ by

$$\bar{v} = \frac{\bar{z}_2}{\bar{z}_1} = \frac{\rho_l^{1/2} v_l + \rho_r^{1/2} v_r}{\rho_l^{1/2} + \rho_r^{1/2}}. \tag{14.54}$$

This is often called the rho-averaged (Roe-averaged?) velocity. We can view $\hat{A}$ as being the Jacobian matrix $f'(u)$ from (14.48) evaluated at the averaged velocity $\bar{v}$. Clearly, when $u_l = u_r$ and $\bar{z} = z_l = z_r$, the matrix $\hat{A}(u_l, u_l)$ reduces to $f'(u_l)$. Since the coefficients of $\hat{A}$ are smooth functions of u_l and u_r, this shows that (14.19iii) is satisfied. Finally, condition (14.19ii) is also clearly satisfied since the eigenvalues and eigenvectors of $\hat{A}$ are given by:

$$\hat{\lambda}_1 = \bar{v} - a, \qquad \hat{\lambda}_2 = \bar{v} + a \tag{14.55}$$

$$\hat{r}_1 = \begin{bmatrix} 1 \\ \bar{v} - a \end{bmatrix}, \qquad \hat{r}_2 = \begin{bmatrix} 1 \\ \bar{v} + a \end{bmatrix}. \tag{14.56}$$

Again, note the similarity to the expressions (7.6) and (7.7) for the eigenvalues and eigenvectors of $f'(u)$.

The approximate solution to the Riemann problem with data u_l and u_r is thus

$$\hat{u}(x, t) = \begin{cases} u_l & x/t < \bar{v} - a \\ \hat{u}_m & \bar{v} - a < x/t < \bar{v} + a \\ u_r & x/t > \bar{v} + a \end{cases} \tag{14.57}$$

where $\hat{u}_m$ is the intermediate state. We can obtain this state by decomposing

$$u_r - u_l = \alpha_1 \hat{r}_1 + \alpha_2 \hat{r}_2 \tag{14.58}$$

and then

$$\hat{u}_m = u_l + \alpha_1 \hat{r}_1. \tag{14.59}$$

EXERCISE 14.4. *Find a Roe matrix for the one-dimensional shallow water equations (5.38).*

15 Nonlinear Stability

15.1 Convergence notions

The Lax-Wendroff Theorem presented in Chapter 12 does not say anything about whether the method converges, only that if a sequence of approximations converges then the limit is a weak solution. To guarantee convergence, we need some form of stability, just as for linear problems. Unfortunately, the Lax Equivalence Theorem no longer holds and we cannot use the same approach (which relies heavily on linearity) to prove convergence. In this chapter we consider one form of nonlinear stability that allows us to prove convergence results for a wide class of practical methods. So far, this approach has been completely successful only for scalar problems. For general systems of equations with arbitrary initial data no numerical method has been proved to be stable or convergent, although convergence results have been obtained in some special cases (e.g. [20], [50], [53]).

You may wonder why we should bother carefully studying the scalar case, since it has limited direct applicability to real-world problems. However, the fact is that many of the most successful numerical methods for systems like the Euler equations have been developed by first inventing good methods for the scalar case (where the theory provides good guidance) and then extending them in a relatively straight forward way to systems of equations. The fact that we can prove they work well for scalar equations is no guarantee that they will work at all for systems, but in practice this approach has been very successful.

One difficulty immediately presents itself when we contemplate the convergence of a numerical method for conservation laws. The global error $U_k(x,t) - u(x,t)$ is not well defined when the weak solution u is not unique. Instead, we measure the global error in our approximation by the distance from $U_k(x,t)$ to the set of *all* weak solutions $\mathcal{W}$,

$$\mathcal{W} = \{w : w(x,t) \text{ is a weak solution to the conservation law}\}. \tag{15.1}$$

To measure this distance we need a norm, for example the 1-norm over some finite time interval $[0, T]$, denoted by

$$\|v\|_{1,T} = \int_0^T \|v(\cdot, t)\|_1 \, dt \tag{15.2}$$

$$= \int_0^T \int_{-\infty}^{\infty} |v(x,t)| \, dx \, dt.$$

The global error is then defined by

$$\text{dist}(U_k, \mathcal{W}) = \inf_{w \in \mathcal{W}} \|U_k - w\|_{1,T}. \tag{15.3}$$

The convergence result we would now like to prove takes the following form:

> If U_k is generated by a numerical method in conservation form, consistent with
> the conservation law, and if the method is stable in some appropriate sense,
> then
> $$\text{dist}(U_k, \mathcal{W}) \to 0 \quad \text{as} \quad k \to 0.$$

Note that there is no guarantee that $\|U_k - w\|_{1,T} \to 0$ as $k \to 0$ for any fixed weak solution $w(x, t)$. The computed U_k might be close to one weak solution for one value of k and close to a completely different weak solution for a slightly smaller value of the time step k. (Recall Exercise 12.4.) This is of no great concern, since in practice we typically compute only on one particular grid, not a sequence of grids with $k \to 0$, and what the convergence result tells us is that by taking a fine enough grid, we can be guaranteed of being arbitrarily close to *some* weak solution.

Of course in situations where there is a unique physically relevant weak solution satisfying some entropy condition, we would ultimately like to prove convergence to this particular weak solution. This issue will be discussed later, since it requires some discrete entropy condition as well as consistency and stability, and is a stronger form of convergence than we now consider.

15.2 Compactness

In order to prove a convergence result of the type formulated above, we must define an appropriate notion of "stability". For nonlinear problems the primary tool used to prove convergence is *compactness*, and so we will take a slight detour to define this concept and indicate its use.

There are several equivalent definitions of a compact set within some normed space. Here I will simply assert that certain classes of sets are compact without proof, since these are standard examples from real or functional analysis and a complete discussion would take us too far afield.

The most important property of compact sets, in relation to our goals of defining stability and proving convergence, is the following.

PROPOSITION 15.1. *If K is a compact set in some normed space, then any infinite sequence of elements of K, $\{\kappa_1,\ \kappa_2,\ \kappa_3,\ ...\}$, contains a subsequence which converges to an element of K.*

This means that from the original sequence we can, by selecting certain elements from this sequence, construct a new infinite sequence

$$\{\kappa_{i_1},\ \kappa_{i_2},\ \kappa_{i_3},\ ...\} \qquad (\text{with } i_1 < i_2 < i_3 < \cdots)$$

which converges to some element $\kappa \in K$,

$$\|\kappa_{i_j} - \kappa\| \to 0 \qquad \text{as } j \to \infty.$$

The fact that compactness guarantees the existence of convergent subsequences, combined with the Lax-Wendroff Theorem 12.1, will give us a convergence proof of the type formulated above.

EXAMPLE 15.1.   In the space $\mathbb{R}$ with norm given by the absolute value, any closed interval is a compact set. So, for example, any sequence of real numbers in $[0, 1]$ contains a subsequence which converges to a number between 0 and 1. Of course there may be several different subsequences one could extract, converging perhaps to different numbers. For example, the sequence

$$\{0, 1, 0, 1, 0, 1, ...\}$$

contains subsequences converging to 0 and subsequences converging to 1.

EXAMPLE 15.2.   In the same space as the previous example, an open interval is *not* compact. For example, the sequence

$$\{1,\ 10^{-1},\ 10^{-2},\ 10^{-3},\ ...\}$$

of elements lying in the open interval $(0, 1)$ contains no subsequences convergent to an element of $(0, 1)$. (Of course the whole sequence, and hence every subsequence, converges to 0, but this number is not in $(0,1)$.) Also, an unbounded set, e.g. $[0, \infty)$, is not compact, since the sequence $\{1,\ 2,\ 3,\ ...\}$ contains no convergent subsequence.

Finite dimensional spaces.   Generalizing the first example above, in any finite dimensional normed linear space, any *closed* and *bounded* set is *compact*. (In fact, these are the only compact sets.)

EXAMPLE 15.3.   In the n-dimensional space $\mathbb{R}^n$ with any vector norm $\|\cdot\|$, the closed ball

$$B_R = \{x \in \mathbb{R}^n :\ \|x\| \le R\}$$

is a compact set.

Function spaces. Since we are interested in proving the convergence of a sequence of functions $U_k(x, t)$, our definition of stability will require that all the functions lie within some compact set in some normed *function space*. Restricting our attention to the time interval $[0, T]$, the natural function space is the space $L_{1,T}$ consisting of all functions of x and t for which the $1, T$-norm (15.2) is finite,

$$L_{1,T} = \{v : \|v\|_{1,T} < \infty\}.$$

This is an infinite dimensional space and so it is not immediately clear what comprises a compact set in this space. Recall that the dimension of a linear space is the number of elements in a basis for the space, and that a *basis* is a set of linearly independent elements with the property that any element can be expressed as a linear combination of the basis elements. Any space with n linearly independent elements has dimension at least n.

EXAMPLE 15.4. The space of functions of x alone with finite 1-norm is denoted by L_1,

$$L_1 = \{v(x) : \|v\|_1 < \infty\}.$$

This space is clearly infinite dimensional since the infinite set of functions

$$v_j(x) = \begin{cases} 1 & j < x < j + 1 \\ 0 & \text{otherwise} \end{cases} \tag{15.4}$$

are linearly independent.

Unfortunately, in an infinite dimensional space, a closed and bounded set is not necessarily compact.

EXAMPLE 15.5. The sequence of function $\{v_1, v_2, \ldots\}$ with v_j defined by (15.4) all lie in the closed and bounded unit ball

$$B_1 = \{v \in L_1 : \|v\|_1 \leq 1\},$$

and yet this sequence has no convergent subsequences.

The difficulty here is that the support of these functions is nonoverlapping and marches off to infinity as $j \to \infty$. We might try to avoid this by considering a set of the form

$$\{v \in L_1 : \|v\|_1 \leq R \text{ and } \text{Supp}(v) \subset [-M, M]\}$$

for some $R, M > 0$, where $\text{Supp}(v)$ denotes the support of the function v, i.e., $\text{Supp}(v) \subset [-M, M]$ means that $v(x) \equiv 0$ for $|x| > M$. However, this set is also not compact, as shown by the sequence of functions $\{v_1, v_2, \ldots\}$ with

$$v_j(x) = \begin{cases} \sin(jx) & |x| \leq 1 \\ 0 & |x| > 1. \end{cases}$$

Again this sequence has no convergent subsequences, now because the functions become more and more oscillatory as $j \to \infty$.

15.3 Total variation stability

In order to obtain a compact set in L_1, we will put a bound on the total variation of the functions, a quantity already defined in (12.40) through (12.42). The set

$$\{v \in L_1 : TV(v) \leq R \text{ and } \text{Supp}(v) \subset [-M, M]\} \tag{15.5}$$

is a compact set, and any sequence of functions with uniformly bounded total variation and support must contain convergent subsequences. (Note that the 1-norm will also be uniformly bounded as a result, with $\|v\|_1 \leq MR$.)

Since our numerical approximations U_k are functions of x and t, we need to bound the total variation in both space and time. We define the total variation over $[0, T]$ by

$$\begin{aligned}
TV_T(u) &= \limsup_{\epsilon \to 0} \frac{1}{\epsilon} \int_0^T \int_{-\infty}^{\infty} |u(x + \epsilon, t) - u(x, t)| \, dx \, dt \\
&+ \limsup_{\epsilon \to 0} \frac{1}{\epsilon} \int_0^T \int_{-\infty}^{\infty} |u(x, t + \epsilon) - u(x, t)| \, dx \, dt.
\end{aligned} \tag{15.6}$$

It can be shown that the set

$$\mathcal{K} = \{u \in L_{1,T} : TV_T(u) \leq R \text{ and } \text{Supp}(u(\cdot, t)) \subset [-M, M] \, \forall t \in [0, T]\} \tag{15.7}$$

is a compact set in $L_{1,T}$.

Since our functions $U_k(x, t)$ are always piecewise constant, the definition (15.6) of TV_T reduces to simply

$$TV_T(U^n) = \sum_{n=0}^{T/k} \sum_{j=-\infty}^{\infty} \left[k|U_{j+1}^n - U_j^n| + h|U_j^{n+1} - U_j^n| \right]. \tag{15.8}$$

Note that we can rewrite this in terms of the one-dimensional total variation and 1-norm as

$$TV_T(U^n) = \sum_{n=0}^{T/k} \left[k\, TV(U^n) + \|U^{n+1} - U^n\|_1 \right]. \tag{15.9}$$

DEFINITION 15.1. *We will say that a numerical method is total variation stable, or simply* **TV-stable**, *if all the approximations U_k for $k < k_0$ lie in some fixed set of the form (15.7) (where R and M may depend on the initial data u_0 and the flux function $f(u)$, but not on k).*

Note that our requirement in (15.7) that $\text{Supp}(u)$ be uniformly bounded over $[0, T]$ is always satisfied for any explicit method if the initial data u_0 has compact support and

k/h is constant as $k \to 0$. This follows from the finite speed of propagation under such a method.

The other requirement for TV-stability can be simplified considerably by noting the following theorem. This says that for the special case of functions generated by conservative numerical methods, it suffices to insure that the *one-dimensional* total variation at each time t_n is uniformly bounded (independent of n). Uniform boundedness of TV_T then follows.

THEOREM 15.1. *Consider a conservative method with a Lipschitz continuous numerical flux $F(U; j)$ and suppose that for each initial data u_0 there exists some $k_0, R > 0$ such that*

$$TV(U^n) \le R \qquad \forall\, n, k \text{ with } k < k_0, \quad nk \le T. \qquad (15.10)$$

Then the method is TV-stable.

To prove this theorem we use the following lemma.

LEMMA 15.1. *The bound (15.10) implies that there exists $\alpha > 0$ such that*

$$\|U^{n+1} - U^n\|_1 \le \alpha k \qquad \forall\, n, k \text{ with } k < k_0, \quad nk \le T. \qquad (15.11)$$

PROOF OF THEOREM 15.1. Using (15.10) and (15.11) in (15.9) gives

$$
\begin{aligned}
TV_T(U^n) &= \sum_{n=0}^{T/k} \left[k\, TV(U^n) + \|U^{n+1} - U^n\|_1 \right] \\
&\le \sum_{n=0}^{T/k} [kR + \alpha k] \\
&\le k(R + \alpha)T/k = (R + \alpha)T
\end{aligned}
$$

for all $k < k_0$, showing that $TV_T(U_k)$ is uniformly bounded as $k \to 0$. This, together with the finite speed of propagation argument outlined above, shows that all U_k lie in a set of the form (15.7) for all $k < k_0$ and the method is TV-stable. ∎

PROOF OF LEMMA 15.1. Recall that a method in conservation form has

$$U_j^{n+1} - U_j^n = \frac{k}{h}[F(U^n; j) - F(U^n; j-1)]$$

and hence

$$\|U^{n+1} - U^n\|_1 = k \sum_{j=-\infty}^{\infty} |F(U^n; j) - F(U^n; j-1)|. \qquad (15.12)$$

The flux $F(U; j)$ depends on a finite number of values $U_{j-p}, \ldots, U_{j+q}$. The bound (15.10) together with the compact support of each U^n easily gives

$$|U_j^n| \le R/2 \qquad \forall\, j, n \text{ with } nk \le T. \qquad (15.13)$$

This uniform bound on U_j^n, together with the continuity of $F(U; j)$ and its Lipschitz continuity, allows us to derive a bound of the form

$$|F(U^n; j) - F(U^n; j - 1)| \leq K \max_{-p \leq i \leq q} |U_{j+i}^n - U_{j+i-1}^n|. \tag{15.14}$$

It follows that

$$|F(U^n; j) - F(U^n; j - 1)| \leq K \sum_{i=-p}^{q} |U_{j+i}^n - U_{j+i-1}^n|$$

and so (15.12) gives

$$\|U^{n+1} - U^n\|_1 \leq kK \sum_{i=-p}^{q} \sum_{j=-\infty}^{\infty} |U_{j+i}^n - U_{j+i-1}^n|$$

after interchanging sums. But now the latter sum is simply $TV(U^n)$ for any value of i, and so

$$\|U^{n+1} - U^n\|_1 \ \leq \ kK \sum_{i=-p}^{q} TV(U^n)$$
$$\leq \ kK(p + q + 1)R$$

yielding the bound (15.11). ∎

We are now set to prove our convergence theorem, which requires total variation stability along with consistency.

THEOREM 15.2. *Suppose U_k is generated by a numerical method in conservation form with a Lipschitz continuous numerical flux, consistent with some scalar conservation law. If the method is TV-stable, i.e., if $TV(U^n)$ is uniformly bounded for all n, k with $k < k_0$, $nk \leq T$, then the method is convergent, i.e., $\text{dist}(U_k, \mathcal{W}) \to 0$ as $k \to 0$.*

PROOF. To prove this theorem we suppose that the conclusion is false, and obtain a contradiction. If $\text{dist}(U_k, \mathcal{W})$ does not converge to zero, then there must be some $\epsilon > 0$ and some sequence of approximations $\{U_{k_1}, U_{k_2}, \ldots\}$ such that $k_j \to 0$ as $j \to \infty$ while

$$\text{dist}(U_{k_j}, \mathcal{W}) > \epsilon \quad \text{for all } j. \tag{15.15}$$

Since $U_{k_j} \in \mathcal{K}$ (the compact set of (15.7)) for all j, this sequence must have a convergent subsequence, converging to some function $v \in \mathcal{K}$. Hence far enough out in this subsequence, U_{k_j} must satisfy

$$\|U_{k_j} - v\|_{1,T} < \epsilon \quad \text{for all } j \text{ sufficiently large} \tag{15.16}$$

for the ϵ defined above. Moreover, since the U_k are generated by a conservative and consistent method, it follows from the Lax-Wendroff Theorem (Theorem 12.1) that the limit v must be a weak solution of the conservation law, *i.e.*, $v \in \mathcal{W}$. But then (15.16) contradicts (15.15) and hence a sequence satisfying (15.15) cannot exist and we conclude that $\mathrm{dist}(U_k, \mathcal{W}) \to 0$ as $k \to 0$. $\blacksquare$

15.4   Total variation diminishing methods

We have just seen that TV-stability of a consistent and conservative numerical method is enough to guarantee convergence, in the sense that $\mathrm{dist}(U_k, \mathcal{W}) \to 0$ as $k \to 0$.

One easy way to ensure TV-stability is to require that the total variation be nonincreasing as time evolves, so that the total variation at any time is uniformly bounded by the total variation of the initial data. This requirement gives rise to the very important class of TVD methods.

DEFINITION 15.2. *The numerical method* $U_j^{n+1} = \mathcal{H}(U^n; j)$ *is called* **total variation diminishing** *(abbreviated* **TVD***) if*

$$TV(U^{n+1}) \le TV(U^n) \tag{15.17}$$

for all grid functions U^n.

If we compute using a TVD method, then

$$TV(U^n) \le TV(U^0) \le TV(u_0) \tag{15.18}$$

for all $n \ge 0$.

It can be shown that the true solution to a scalar conservation law has this TVD property, *i.e.*, any weak solution $u(x, t)$ satisfies

$$TV(u(\cdot, t_2)) \le TV(u(\cdot, t_1)) \qquad \text{for } t_2 \ge t_1. \tag{15.19}$$

If this were not the case then it would be impossible to develop a TVD numerical method. However, since true solutions are TVD, it is reasonable to impose this requirement on the numerical solution as well, yielding a TV-stable and hence convergent method. A number of very successful numerical methods have been developed using this requirement. Several such methods will be derived in Chapter 16.

15.5   Monotonicity preserving methods

Recall that one difficulty associated with numerical approximations of discontinuous solutions is that oscillations may appear near the discontinuity. In an attempt to eliminate this possibility, one natural requirement we might place on a numerical method is that it be **monotonicity preserving**. This means that if the initial data U_j^0 is monotone (either nonincreasing or nondecreasing) as a function of j, then the solution U_j^n should have the same property for all n. For example, if $U_j^0 \geq U_{j+1}^0$ for all j then we should have that $U_j^n \geq U_{j+1}^n$ for all j and n (as for example in Figure 12.2). This means in particular that oscillations cannot arise near an isolated propagating discontinuity, since the Riemann initial data is monotone. For TVD methods we have this property.

THEOREM 15.3. *Any TVD method is monotonicity preserving.*

PROOF. This result follows from the fact that the appearance of oscillations would increase the total variation. If $U_j^0 \geq U_{j+1}^0$ for all j and $TV(U^0) < \infty$ then we must have that $TV(U^0) = |U_{-\infty}^0 - U_\infty^0|$. Clearly, by finite domain of dependence arguments, we continue to have $U_j^n \to U_{\pm\infty}^0$ as $j \to \pm\infty$ at any later time t_n. So $TV(U^n) \geq |U_{-\infty}^0 - U_\infty^0|$. If the method is TVD then it must be that in fact $TV(U^n) = |U_{-\infty}^0 - U_\infty^0|$. Any oscillations in U^n would give a larger total variation and cannot occur. ∎

Another attractive feature of the TVD requirement is that it is possible to derive methods with a high order of accuracy which are TVD. By contrast, if we define "stability" by mimicing certain other properties of the true solution, we find that accuracy is limited to first order. Nonetheless, we introduce some of these other concepts because they are useful and frequently seen.

15.6 l_1-contracting numerical methods

Any weak solution of a scalar conservation law satisfies

$$\|u(\cdot, t_2)\|_1 \leq \|u(\cdot, t_1)\|_1 \qquad \text{for } t_2 \geq t_1. \tag{15.20}$$

In particular, if u_0 is the initial data at time $t = 0$, then

$$\|u(\cdot, t)\|_1 \leq \|u_0\|_1 \qquad \forall t \geq 0. \tag{15.21}$$

The result (15.20) is true for any weak solution to a scalar conservation law. If we restrict our attention to the entropy solution (which is unique) then (15.20) is a special case of a more general result. If $u(x,t)$ and $v(x,t)$ are both entropy solutions of the same scalar conservation law (but with possibly different initial data), and if $u_0 - v_0$ has compact support (so that $\|u(\cdot, t) - v(\cdot, t)\|_1 < \infty$ for all t) then

$$\|u(\cdot, t_2) - v(\cdot, t_2)\|_1 \leq \|u(\cdot, t_1) - v(\cdot, t_1)\|_1 \qquad \text{for } t_2 \geq t_1. \tag{15.22}$$

This property is called L_1-**contraction**: $u - v$ is contracting in the 1-norm as time evolves. We can derive (15.20) from (15.22) by taking $v(x, t) \equiv 0$, which is an entropy solution to any conservation law.

Note that we must restrict our attention to the unique entropy solution in order to expect that (15.22) holds. Otherwise, we could take data $u(x, t_1) = v(x, t_1)$ for which two different weak solutions $u(x, t_2)$ and $v(x, t_2)$ exist, violating (15.22).

The standard proof of (15.22) proceeds by defining a finite difference equation satisfying a discrete analogue of (15.22) and then proving convergence (see, e.g., Theorem 16.1 in [77]). This suggests that a discrete analogue of (15.22) could be useful in proving convergence of difference methods.

The discrete space l_1. For grid functions $U = \{U_j\}$ (at a fixed time, for example) we define the 1-norm by

$$\|U\|_1 = h \sum_{j=-\infty}^{\infty} |U_j|. \tag{15.23}$$

The space l_1 consists of all grid functions for which the 1-norm is finite:

$$l_1 = \{U : \|U\|_1 < \infty\}. \tag{15.24}$$

Note that if we extend the grid function U to a piecewise constant function $\tilde{u}(x) = U_j$ for $x_{j-1/2} \le x < x_{j+1/2}$, then

$$\|U\|_1 = \|\tilde{u}\|_1. \tag{15.25}$$

Conversely, if we take a function $u(x)$ and restrict it to a grid function U by setting $U_j = \bar{u}_j$, the cell average (10.3), then

$$\|U\|_1 \le \|u\|_1. \tag{15.26}$$

EXERCISE 15.1. *Verify (15.25) and (15.26).*

In analogy to the L_1-contraction property (15.22) of the true solution operator, we say that a numerical method

$$U_j^{n+1} = \mathcal{H}(U^n; j) \tag{15.27}$$

is l_1-contracting if, for any two grid functions U^n and V^n for which $U^n - V^n$ has compact support, the grid functions U^{n+1} and V^{n+1} defined by (15.27) and $V_j^{n+1} = \mathcal{H}(V^n; j)$ satisfy

$$\|U^{n+1} - V^{n+1}\|_1 \le \|U^n - V^n\|_1. \tag{15.28}$$

The fact that l_1-contracting methods are convergent follows from the next theorem and our previous results.

THEOREM 15.4. *Any l_1-contracting numerical method is TVD.*

PROOF. The proof depends on the following important relation between the 1-norm and total variation: Given any grid function U, define V by shifting U,

$$V_j = U_{j-1} \quad \forall j. \tag{15.29}$$

Then

$$TV(U) = \frac{1}{h}\|U - V\|_1.$$
(15.30)

Now suppose the method (15.27) is l_1-contracting and define $V_j^n = U_{j-1}^n$ for all j and n. Note that the methods we are considering are translation invariant, so $V_j^{n+1} = \mathcal{H}(V^n; j)$. Then by l_1-contraction and (15.30) we have

$$
\begin{aligned}
TV(U^{n+1}) &= \frac{1}{h}\|U^{n+1} - V^{n+1}\|_1 \\
&\leq \frac{1}{h}\|U^n - V^n\|_1 \\
&= TV(U^n)
\end{aligned}
$$

and hence the method is TVD. ∎

EXAMPLE 15.6. We will show that the upwind method is l_1-contracting and hence TVD, provided the CFL condition is satisfied. Suppose, for simplicity, that $f'(U_j^n) > 0$ and $f'(V_j^n) > 0$ for all j. Then the method for U and V reduces to

$$U_j^{n+1} = U_j^n - \frac{k}{h}\left[f(U_j^n) - f(U_{j-1}^n)\right]$$
(15.31)

and

$$V_j^{n+1} = V_j^n - \frac{k}{h}\left[f(V_j^n) - f(V_{j-1}^n)\right].$$
(15.32)

Since the characteristic wave speed is $f'(u)$, the CFL condition requires that

$$0 \leq \frac{k}{h}f'(u) \leq 1$$
(15.33)

for all u in the range $\min_j(U_j^n, V_j^n) \leq u \leq \max_j(U_j^n, V_j^n)$.

Letting $W_j^n = U_j^n - V_j^n$, we find by subtracting (15.32) from (15.31) that

$$W_j^{n+1} = W_j^n - \frac{k}{h}\left[\left(f(U_j^n) - f(V_j^n)\right) - \left(f(U_{j-1}^n) - f(V_{j-1}^n)\right)\right].$$
(15.34)

Since f is smooth, the mean value theorem says that

$$
\begin{aligned}
f(U_j^n) - f(V_j^n) &= f'(\theta_j^n)(U_j^n - V_j^n) \\
&= f'(\theta_j^n)W_j^n
\end{aligned}
$$

for some θ_j^n between U_j^n and V_j^n. Using this in (15.34) gives

$$W_j^{n+1} = \left(1 - \frac{k}{h}f'(\theta_j^n)\right)W_j^n + \frac{k}{h}f'(\theta_{j-1}^n)W_{j-1}^n.$$

By the CFL condition (since θ_j^n falls in the appropriate range), we see that both of the coefficients are nonnegative and so, letting $\alpha_j = (k/h)f'(\theta_j^n)$,

$$|W_j^{n+1}| \le (1 - \alpha_j)|W_j^n| + \alpha_{j-1}|W_{j-1}^n|. \tag{15.35}$$

Summing over j gives the result for the 1-norm:

$$h\sum_j |W_j^{n+1}| \le h\sum_j |W_j^n| - h\sum_j \alpha_j |W_j^n| + h\sum_j \alpha_{j-1}|W_{j-1}^n|.$$

Since $W_j^n = 0$ for $|j|$ sufficiently large, the last two terms cancel out, and we obtain

$$\|W^{n+1}\|_1 \le \|W^n\|_1 \tag{15.36}$$

showing l_1-contraction. (Note the similarity of this proof to the earlier proof in Example 10.2 that the upwind method is stable for the linear advection equation.)

EXERCISE 15.2. *Show that the Lax-Friedrichs method (12.15) is l_1-contracting provided the CFL condition $|kf'(u)/h| \le 1$ is satisfied for all u in the range $\min_j(U_j^n, V_j^n) \le u \le \max_j(U_j^n, V_j^n)$.*

Note that if our numerical method is l_1-contracting, then in particular we can obtain a discrete version of (15.20) by taking $V_j^0 \equiv 0$ (which leads to $V_j^n \equiv 0$ for any consistent method):

$$\|U^n\|_1 \le \|U^0\|_1 \qquad \forall n \ge 0. \tag{15.37}$$

If we define the piecewise constant function $U_k(x,t)$ in the usual way, we also have that

$$\|U_k(\cdot, t)\|_1 \le \|U_k(\cdot, 0)\|_1 \le \|u_0\|_1 \tag{15.38}$$

by using (15.25) and (15.26).

15.7 Monotone methods

Another useful property of the entropy-satisfying weak solution is the following: If we take two sets of initial data u_0 and v_0, with

$$v_0(x) \ge u_0(x) \quad \forall\, x,$$

then the respective entropy solutions $u(x,t)$ and $v(x,t)$ satisfy

$$v(x,t) \ge u(x,t) \quad \forall\, x, t.$$

The numerical method $U_j^{n+1} = \mathcal{H}(U^n; j)$ is called a **monotone method** if the analogous property holds:

$$V_j^n \ge U_j^n \quad \forall j \quad \Longrightarrow \quad V_j^{n+1} \ge U_j^{n+1} \quad \forall j. \tag{15.39}$$

To prove that a method is monotone, it suffices to check that

$$\frac{\partial}{\partial U_i^n} \mathcal{H}(U^n; j) \geq 0 \qquad \text{for all } i, j, U^n. \tag{15.40}$$

This means that if we increase the value of any U_i^n then the value of U_j^{n+1} cannot decrease as a result.

EXAMPLE 15.7. The Lax-Friedrichs method (12.15) is monotone provided the CFL condition is satisfied since

$$\mathcal{H}(U^n; j) = \frac{1}{2}\left(U_{j-1}^n + U_{j+1}^n\right) - \frac{k}{2h}\left(f(U_{j+1}^n) - f(U_{j-1}^n)\right) \tag{15.41}$$

satisfies

$$\frac{\partial}{\partial U_i^n} \mathcal{H}(U^n; j) = \begin{cases} \frac{1}{2}\left(1 + \frac{k}{h}f'(U_{j-1}^n)\right) & \text{if } i = j - 1 \\ \frac{1}{2}\left(1 - \frac{k}{h}f'(U_{j+1}^n)\right) & \text{if } i = j + 1 \\ 0 & \text{otherwise.} \end{cases} \tag{15.42}$$

The CFL condition guarantees that $1 \pm \frac{k}{h}f'(U_i^n) \geq 0$ for all i and so $\partial\mathcal{H}(U^n; j)/\partial U_i^n \geq 0$ for all i, j.

Monotone methods are a subset of l_1-contracting methods, as the following theorem shows.

THEOREM 15.5. *Any monotone method is l_1-contracting.*

Because the condition (15.40) is usually easy to check, as the example above shows, this is a useful theorem. It simplifies considerably the solution of Exercise 15.2 or Example 15.6.

Proofs of Theorem 15.5 may be found in Keyfitz' appendix to [31] or in Crandall and Majda[16]. This latter paper contains a thorough discussion of monotone methods and their properties.

To summarize the relation between the different types of methods considered above, we have:

$$\text{monotone} \quad \Longrightarrow \quad l_1\text{-contracting} \quad \Longrightarrow \quad \text{TVD} \quad \Longrightarrow \quad \text{monotonicity preserving.}$$

Although the monotone requirement (15.40) is often the easiest to check, the class of monotone methods is greatly restricted as the following theorem shows.

THEOREM 15.6. *A monotone method is at most first order accurate.*

The proof of Theorem 15.6 can be found in [31], and relies on the "modified equation" for the monotone method. As described in Chapter 11, by considering the local truncation error and expanding in Taylor series, it can be shown that the numerical solution is in fact a second order accurate approximation to the solution $v(x, t)$ to some modified PDE

$$v_t + f(v)_x = k\frac{\partial}{\partial x}\left(\beta(v)v_x\right) \tag{15.43}$$

where k is the time step and $\beta(v)$ is a function of v that depends on the derivatives of $\mathcal{H}$ with respect to each argument. The assumption that the method is monotone can be used to show that $\beta(v) > 0$ (except in certain trivial cases). But the equation (15.43) is an $O(k)$ perturbation of the original conservation law and it follows that the solution $v^k(x, t)$ differs from $u(x, t)$ by $O(k)$ as well, at least for nontrivial initial data, and so

$$\|v^k - u\| \geq C_1 k \qquad \forall\, k < k_0$$

with $C_1 > 0$. Since our numerical solution is a second order accurate approximation to the modified equation,

$$\|U_k - v_k\| \leq C_2 k^2 \qquad \forall\, k < k_0.$$

It follows that it can only be first order accurate solution of the original conservation law, since

$$\begin{aligned}
\|U_k - u\| &\geq \|u - v^k\| - \|U_k - v^k\| \\
&\geq C_1 k - C_2 k^2
\end{aligned}$$

and hence the error is asymptotically bounded below by $(C_1 - \epsilon)k$ for some fixed ϵ as $k \to 0$.

The notion of total-variation stability is much more useful because it is possible to derive TVD methods that have better than first order accuracy. This has been a prominent theme in the development of high resolution methods, as we will see in the next chapter.

Numerical viscosity and the entropy condition. Note that the modified equation (15.43) contains a dissipative term $k\frac{\partial}{\partial x}(\beta(v)v_x)$, as in the modified equation for first order methods applied to the linear advection equation, discussed in Section 11.1.1. This is similar to the viscous term ϵv_{xx} we added to our conservation law to define the "vanishing viscosity solution". The viscous term in (15.43) vanishes in the limit as $k \to 0$. This suggests that we can hope for more than mere stability of monotone methods. As the grid is refined, we will in fact have convergence of any sequence of approximations to the vanishing viscosity solution of the conservation law, as the next theorem states.

THEOREM 15.7. *The numerical solution computed with a consistent monotone method with k/h fixed converges to the entropy solution as $k \to 0$.*

Although I have motivated this by the presence of numerical viscosity in the method, the standard proof uses the fact that only the entropy solution u is L_1-contracting in the sense that

$$\|u(\cdot, t_2) - w(\cdot, t_2)\|_1 \leq \|u(\cdot, t_1) - w(\cdot, t_1)\|_1 \qquad \text{for } t_2 \geq t_1 \qquad (15.44)$$

for all *smooth* solutions w. This, combined with Theorem 15.5, gives the proof (see [16] or [31]).

Note that there is no explicit reference to stability bounds on k/h in Theorem 15.7, but that the method will only be monotone for sufficiently small values of k/h (as in

Example 15.7). The monotone property is a form of stability, but one that is too stringent for most practical purposes because of Theorem 15.6. In the next chapter we begin our study of higher order, total variation stable methods.

16 High Resolution Methods

In the previous chapter, we observed that monotone methods for scalar conservation laws are TVD and satisfy a discrete entropy condition. Hence they converge in a nonoscillatory manner to the unique entropy solution. However, monotone methods are at most first order accurate, giving poor accuracy in smooth regions of the flow. Moreover, shocks tend to be heavily smeared and poorly resolved on the grid. These effects are due to the large amount of numerical dissipation in monotone methods. Some dissipation is obviously needed to give nonoscillatory shocks and to ensure that we converge to the vanishing viscosity solution, but monotone methods go overboard in this direction.

In this chapter we will study some "high resolution" methods. This term applies to methods that are at least second order accurate on smooth solutions and yet give well resolved, nonoscillatory discontinuities. We will first consider the scalar problem, where we can impose the constraint that the method be total variation diminishing. This insures that we obtain nonoscillatory shocks and convergence in the sense of Theorem 15.2. These scalar methods will later be extended to systems of equations using an approximate decomposition of the system into characteristic fields.

The main idea behind any high resolution method is to attempt to use a high order method, but to modify the method and increase the amount of numerical dissipation in the neighborhood of a discontinuity.

16.1 Artificial Viscosity

The most obvious way to do this is to simply take a high order method, say Lax-Wendroff, and add an additional "artificial viscosity" term to the hyperbolic equation, perhaps modeling the addition of a diffusive term proportional to u_{xx}. This term must have a coefficient that vanishes as k, $h \to 0$, so that it remains consistent with the hyperbolic equation. Moreover, we want this coefficient to vanish sufficiently quickly that the order of accuracy of the high order method on smooth solutions is unaffected.

Since additional viscosity is typically needed only near discontinuities, we might also want the coefficient to depend on the behavior of the solution, being larger near discontinuities than in smooth regions. We will introduce this complication shortly, but first consider the effect of adding a viscous term that is independent of the solution.

As an example, consider the advection equation $u_t + au_x = 0$ and suppose we modify the standard Lax-Wendroff method from Table 10.1 as follows:

$$U_j^{n+1} = U_j^n - \frac{\nu}{2}(U_{j+1}^n - U_{j-1}^n) + \frac{1}{2}\nu^2(U_{j+1}^n - 2U_j^n + U_{j-1}^n) \qquad (16.1)$$
$$+ kQ(U_{j+1}^n - 2U_j^n + U_{j-1}^n)$$

where $\nu = ak/h$ is the Courant number and Q is the newly-introduced artificial viscosity. Clearly the truncation error $L(x,t)$ for this method can be written in terms of the truncation error $L_{LW}(x,t)$ for the Lax-Wendroff method as

$$
\begin{aligned}
L(x,t) &= L_{LW}(x,t) - Q[u(x+h,t) - 2u(x,t) + u(x-h,t)] \\
&= L_{LW}(x,t) - Qh^2 u_{xx}(x,t) + O(h^4) \\
&= O(k^2) \quad \text{as } k \to 0
\end{aligned}
$$

since $L_{LW}(x,t) = O(k^2)$ and $h^2 = O(k^2)$ as $k \to 0$. The method remains second order accurate for any choice of $Q = $ constant.

The modified equation for (16.1) is similarly related to the modified equation (11.7) for Lax-Wendroff. The method (16.1) produces a third order accurate approximation to the solution of the PDE

$$u_t + au_x = h^2 Q u_{xx} + \frac{1}{6}h^2 a(\nu^2 - 1)u_{xxx}. \qquad (16.2)$$

The dispersive term that causes oscillations in Lax-Wendroff now has competition from a dissipative term, and one might hope that for Q sufficiently large the method would be nonoscillatory. Unfortunately, this is not the case. The method (16.1) with constant Q is still a linear method, and is second order accurate, and so the following theorem due to Godunov shows that it cannot be monotonicity preserving.

THEOREM 16.1 (GODUNOV). *A linear, monotonicity preserving method is at most first order accurate.*

PROOF. We will show that any linear, monotonicity preserving method is monotone, and then the result follows from Theorem 15.6.

Let U^n be any grid function and let $V_j^n = U_j^n$ for $j \neq J$ while $V_J^n > U_J^n$. Then we need to show that $V_j^{n+1} \geq U_j^{n+1}$ for all j, which implies that the method is monotone.

Let W^n be the monotone Riemann data defined by

$$W_j^n = \begin{cases} U_j^n & j < J \\ V_j^n & j \geq J \end{cases} \qquad (16.3)$$

so that W_j^n is nondecreasing. Note that

$$W_j^n = W_{j-1}^n + (V_j^n - U_j^n) \qquad (16.4)$$

for all j, since the last term is zero except when $j = J$. Since the method is linear, we also have from (16.4) that

$$W_j^{n+1} = W_{j-1}^{n+1} + (V_j^{n+1} - U_j^{n+1})$$
(16.5)

so that

$$V_j^{n+1} = U_j^{n+1} + (W_j^{n+1} - W_{j-1}^{n+1}).$$
(16.6)

But since the method is monotonicity preserving and W^n is monotone with $W_{j-1}^n \leq W_j^n$, we have that $W_j^{n+1} - W_{j-1}^{n+1} \geq 0$ and so (16.6) gives $V_j^{n+1} \geq U_j^{n+1}$. This shows that the method is monotone and hence first order. ∎

Nonlinear methods. To have any hope of achieving a monotonicity preserving method of the form (16.1), we must let Q depend on the data U^n so that the method is nonlinear, even for the linear advection equation. As already noted, allowing this dependence makes sense from the standpoint of accuracy as well: where the solution is smooth adding more dissipation merely increases the truncation error and should be avoided. Near a discontinuity we should increase Q to maintain monotonicity (and also, one hopes, enforce the entropy condition).

The idea of adding a variable amount of artificial viscosity, based on the structure of the data, goes back to some of the earliest work on the numerical solution of fluid dynamics equations, notably the paper of von Neumann and Richtmyer[95]. Lax and Wendroff[46] also suggested this in their original presentation of the Lax-Wendroff method.

Note that in order to maintain conservation form, one should introduce variable viscosity by replacing the final term of (16.1) by a term of the form

$$k \left[Q(U^n; j)(U_{j+1}^n - U_j^n) - Q(U^n; j-1)(U_j^n - U_{j-1}^n) \right]$$
(16.7)

where $Q(U^n; j)$ is the artificial viscosity which now depends on some finite set of values of U^n, say $U_{j-p}^n, \ldots, U_{j+q}^n$. More generally, given any high order flux function $F_H(U; j)$ for a general conservation law, the addition of artificial viscosity replaces this by the modified flux function

$$F(U; j) = F_H(U; j) - hQ(U; j)(U_{j+1} - U_j).$$
(16.8)

The difficulty with the artificial viscosity approach is that it is hard to determine an appropriate form for Q that introduces just enough dissipation to preserve monotonicity without causing unnecessary smearing. Typically these goals are not achieved very reliably.

For this reason, the high resolution methods developed more recently are based on very different approaches, in which the nonoscillatory requirement can be imposed more directly. It is generally possible to rewrite the resulting method as a high order method plus some artificial viscosity, but the resulting viscosity coefficient is typically very complicated and not at all intuitive.

There are a wide variety of approaches that can be taken, and often there are close connections between the methods developed by quite different means. We will concentrate on just two classes of methods that are quite popular: flux-limiter methods and slope-limiter methods. More comprehensive reviews of many high resolution methods are given by Colella and Woodward[98] and Zalesak[102].

16.2   Flux-limiter methods

In this approach, we begin by choosing a high order flux $F_H(U;j)$ (e.g., the Lax-Wendroff flux) that works well in smooth regions, and a low order flux $F_L(U;j)$ (typically some monotone method) that behaves well near discontinuities. We then attempt to hybridize these two into a single flux F in such a way that F reduces to F_H in smooth regions and to F_L near discontinuities. The main idea is outlined here, although for nonlinear problems the actual hybridization may be more complicated.

We can view the high order flux as consisting of the low order flux plus a correction:

$$F_H(U;j) = F_L(U;j) + [F_H(U;j) - F_L(U;j)] . \qquad (16.9)$$

In a flux-limiter method, the magnitude of this correction is limited depending on the data, so the flux becomes

$$F(U;j) = F_L(U;j) + \Phi(U;j) [F_H(U;j) - F_L(U;j)] \qquad (16.10)$$

where $\Phi(U;j)$ is the limiter. If the data U is smooth near U_j then $\Phi(U;j)$ should be near 1 while in the vicinity of a discontinuity we want $\Phi(U;j)$ to be near zero. (In practice, allowing a wider range of values for Φ often works better.)

Note that we can rewrite (16.10) as

$$F(U;j) = F_H(U;j) - (1 - \Phi(U;j))[F_H(U;j) - F_L(U;j)] \qquad (16.11)$$

and comparison with (16.8) gives the equivalent artificial viscosity for this type of method.

One of the earliest high resolution methods, the **flux-corrected transport (FCT)** method of Boris and Book[2], can be viewed as a flux-limiter method. They refer to the correction term in (16.9) as the *antidiffusive flux*, since the low order flux F_L contains too much diffusion for smooth data and the correction compensates. The FCT strategy is to add in as much of this antidiffusive flux as possible without increasing the variation of the solution, leading to a simple and effective algorithm.

Hybrid methods of this form were also introduced by Harten and Zwas[36] at roughly the same time. More recently, a wide variety of methods of this form have been proposed, e.g., [27], [60], [65], [101].

A reasonably large class of flux-limiter methods has been studied by Sweby[83], who derived algebraic conditions on the limiter function which guarantee second order accuracy and the TVD property. The discussion here closely follows his presentation.

To introduce the main ideas in a simple setting we first consider the linear advection equation and take F_H to be the Lax-Wendroff flux while F_L is the first order upwind flux. If we assume $a > 0$, then we can rewrite Lax-Wendroff to look like the upwind method plus a correction as follows:

$$U_j^{n+1} = U_j^n - \nu(U_j^n - U_{j-1}^n) - \frac{1}{2}\nu(1-\nu)(U_{j+1}^n - 2U_j^n + U_{j-1}^n). \tag{16.12}$$

The corresponding flux can be written as

$$F(U;j) = aU_j + \frac{1}{2}a(1-\nu)(U_{j+1} - U_j). \tag{16.13}$$

The first term is the upwind flux and the second term is the Lax-Wendroff correction, so that this gives a splitting of the flux of the form (16.9).

To define a flux-limiter method, we replace (16.13) by

$$F(U;j) = aU_j + \frac{1}{2}a(1-\nu)(U_{j+1} - U_j)\phi_j \tag{16.14}$$

where ϕ_j is shorthand for $\Phi(U;j)$, and represents the flux-limiter.

There are various ways one might measure the "smoothness" of the data. One possibility is to look at the ratio of consecutive gradients,

$$\theta_j = \frac{U_j - U_{j-1}}{U_{j+1} - U_j}. \tag{16.15}$$

If θ_j is near 1 then the data is presumably smooth near U_j. If θ_j is far from 1 then there is some sort of kink in the data at U_j. We can then take ϕ_j to be a function of θ_j,

$$\phi_j = \phi(\theta_j) \tag{16.16}$$

where ϕ is some given function.

Note that this measure of smoothness breaks down near extreme points of U, where the denominator may be close to zero and θ_j arbitrarily large, or negative, even if the solution is smooth. As we will see, maintaining second order accuracy at extreme points is impossible with TVD methods. For the time being, we will be content with second order accuracy away from these points, and the following theorem gives conditions on ϕ which guarantee this.

THEOREM 16.2. *The flux limiter method with flux (16.14) (where ϕ_j is given by (16.16)) is consistent with the advection equation provided $\phi(\theta)$ is a bounded function. It* is second order accurate (on smooth solutions with u_x bounded away from zero) provided $\phi(1) = 1$ with ϕ Lipschitz continuous at $\theta = 1$.

EXERCISE 16.1. *Prove this theorem.*

To see what conditions are required to give a TVD method, we use (16.14) to obtain the following method (dropping the superscripts on U^n for clarity):

$$U_j^{n+1} = U_j - \frac{k}{h}\left\{a(U_j - U_{j-1}) + \frac{1}{2}a(1-\nu)\left[(U_{j+1} - U_j)\phi_j - (U_j - U_{j-1})\phi_{j-1}\right]\right\}$$
$$= U_j - \left(\nu - \frac{1}{2}\nu(1-\nu)\phi_{j-1}\right)(U_j - U_{j-1}) - \frac{1}{2}\nu(1-\nu)\phi_j(U_{j+1} - U_j). \quad (16.17)$$

If we consider a general method of the form

$$U_j^{n+1} = U_j - C_{j-1}(U_j - U_{j-1}) + D_j(U_{j+1} - U_j), \quad\quad\quad (16.18)$$

then the following theorem of Harten[27] can be used to give constraints on the ϕ_j.

THEOREM 16.3 (HARTEN). *In order for the method (16.18) to be TVD, the following conditions on the coefficients are sufficient:*

$$\begin{aligned} C_{j-1} &\geq 0 \quad \forall j \\ D_j &\geq 0 \quad \forall j \\ C_j + D_j &\leq 1 \quad \forall j \end{aligned} \quad\quad\quad (16.19)$$

PROOF. The resulting method can be shown to be TVD by computing

$$U_{j+1}^{n+1} - U_j^{n+1} = (1 - C_j - D_j)(U_{j+1} - U_j) + D_{j+1}(U_{j+2} - U_{j+1})$$
$$+ C_{j-1}(U_j - U_{j-1}).$$

We now sum $|U_{j+1}^{n+1} - U_j^{n+1}|$ over j and use the nonnegativity of each coefficient as in previous arguments of this type to show that $TV(U^{n+1}) \leq TV(U^n)$. ∎

EXERCISE 16.2. *Complete this proof.*

The form (16.17) suggests that we try taking

$$C_{j-1} = \nu\left(1 - \frac{1}{2}(1-\nu)\phi_{j-1}\right)$$
$$D_j = -\frac{1}{2}\nu(1-\nu)\phi_j.$$

Unfortunately, there is no hope of satisfying the condition (16.19) using this, since $D_j < 0$ when ϕ_j is near 1.

However, we can also obtain (16.17) by taking

$$C_{j-1} = \nu + \frac{1}{2}(1-\nu)\nu \left[\frac{\phi_j(U_{j+1} - U_j) - \phi_{j-1}(U_j - U_{j-1})}{U_j - U_{j-1}} \right],$$

$$D_j = 0.$$

The conditions (16.19) are then satisfied provided

$$0 \le C_{j-1} \le 1. \tag{16.20}$$

Using (16.15) and (16.16), we can rewrite the expression for C_{j-1} as

$$C_{j-1} = \nu \left\{ 1 + \frac{1}{2}(1-\nu) \left[\frac{\phi(\theta_j)}{\theta_j} - \phi(\theta_{j-1}) \right] \right\}. \tag{16.21}$$

The condition (16.20) is satisfied provided the CFL condition $|\nu| \le 1$ holds along with the bound

$$\left| \frac{\phi(\theta_j)}{\theta_j} - \phi(\theta_{j-1}) \right| \le 2 \qquad \text{for all } \theta_j, \ \theta_{j-1}. \tag{16.22}$$

If $\theta_j \le 0$ then the slopes at neighboring points have opposite signs. The data then has an extreme point near U_j and the total variation will certainly increase if the value at this extreme point is accentuated. For this reason, it is safest to take $\phi(\theta) = 0$ for $\theta \le 0$ and use the upwind method alone. Note that this is unsatisfying since if the data is smooth near the extreme point, we would really like to take ϕ near 1 so that the high order method is being used. However, the total variation will generally increase if we do this. Osher and Chakravarthy[60] prove that TVD methods must in fact degenerate to first order accuracy at extreme points.

More recently, it has been shown by Shu[75] that a slight modification of these methods, in which the variation is allowed to increase by $O(k)$ in each time step, can eliminate this difficulty. The methods are no longer TVD but are total variation stable since over a finite time domain uniform bounds on the total variation can be derived. If we have

$$TV(U^{n+1}) \le (1 + \alpha k)TV(U^n), \tag{16.23}$$

where α is independent of U^n, then

$$TV(U^n) \le (1 + \alpha k)^n TV(U^0) \le e^{\alpha T} TV(u_0) \tag{16.24}$$

for $nk \le T$ and hence the method is total variation stable.

For simplicity here, we will only consider TVD methods and assume that

$$\phi(\theta) = 0 \quad \text{for } \theta \le 0. \tag{16.25}$$

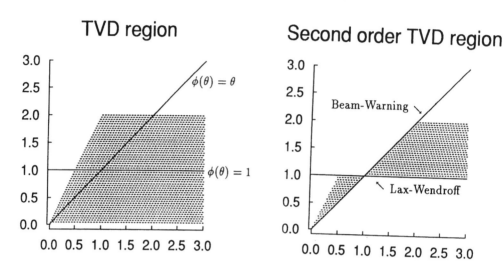

Figure 16.1. Regions in which function values $\phi(\theta)$ must lie in order to give TVD and second order TVD methods.

Then (16.22) will be satisfied provided

$$0 \le \frac{\phi(\theta)}{\theta} \le 2 \quad \text{and} \quad 0 \le \phi(\theta) \le 2 \tag{16.26}$$

for all θ.

This region is shown in Figure 16.1. To obtain second order accuracy, the function ϕ must also pass smoothly through the point $\phi(1) = 1$. Sweby found, moreover, that it is best to take ϕ to be a convex combination of the ϕ for Lax-Wendroff (which is simply $\phi \equiv 1$) and the ϕ for Beam-Warming (which is easily seen to be $\phi(\theta) = \theta$). Other choices apparently give too much compression, and smooth data such as a sine wave tends to turn into a square wave as time evolves. Imposing this additional restriction gives the "second order TVD" region of Sweby which is also shown in Figure 16.1.

EXAMPLE 16.1. If we define $\phi(\theta)$ by the upper boundary of the second order TVD region shown in Figure 16.1, *i.e.*,

$$\phi(\theta) = \max(0, \ \min(1, 2\theta), \ \min(\theta, 2)), \tag{16.27}$$

then we obtain the so-called "superbee" limiter of Roe[67].

A smoother limiter function, used by van Leer[89], is given by

$$\phi(\theta) = \frac{|\theta| + \theta}{1 + |\theta|}. \tag{16.28}$$

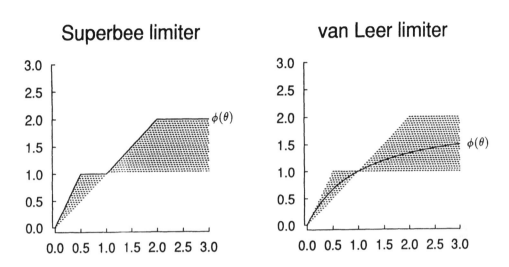

Figure 16.2. *Two possible limiter functions: Roe's "superbee" and van Leer's limiter.*

These limiters are shown in Figure 16.2.

Sweby[83] gives several other examples and presents some numerical comparisons. More extensive comparisons for the linear advection equation are presented in Zalesak[102].

Sweby also discusses the manner in which this approach is extended to nonlinear scalar conservation laws. The basic idea is to replace $\nu = ak/h$ by a "local ν" defined at each grid point by

$$\nu_j = \frac{k}{h} \left(\frac{f(U_{j+1}) - f(U_j)}{U_{j+1} - U_j} \right). \tag{16.29}$$

The resulting formulas are somewhat complicated and will not be presented here. They are similar to the nonlinear generalization of the slope-limiter methods which will be presented below.

Generalization to arbitrary wave speeds. In the above description we assumed $a > 0$. Obviously, a similar method can be defined when $a < 0$ by again viewing Lax-Wendroff as a modification of the upwind method, which is now one-sided in the opposite direction. It is worth noting that we can unify these methods into a single formula. This is useful in generalizing to linear systems and nonlinear problems, where both positive and negative wave speeds can exist simultaneously.

Recall that the upwind method for a linear system can be written in the form (13.15), which in the scalar case reduces to

$$F_L(U; j) = \frac{1}{2} a(U_j + U_{j+1}) - \frac{1}{2} |a|(U_{j+1} - U_j) \tag{16.30}$$

and is now valid for a of either sign. Also notice that the Lax-Wendroff flux can be written as

$$F_H(U;j) = \frac{1}{2}a(U_j + U_{j+1}) - \frac{1}{2}\nu a(U_{j+1} - U_j). \tag{16.31}$$

This can be viewed as a modification of F_L, and introducing a limiter as in (16.10) gives the flux

$$F(U;j) = F_L(U;j) + \frac{1}{2}\phi_j(\mathrm{sgn}(\nu) - \nu)a(U_{j+1} - U_j). \tag{16.32}$$

Note that we have used the fact that $|a| = \mathrm{sgn}(a)a = \mathrm{sgn}(\nu)a$, since $\nu = ak/h$ and $k,\ h > 0$. The flux-limiter ϕ_j is again of the form $\phi_j = \phi(\theta_j)$, but we now take θ_j to be a ratio of slopes in the upwind direction, which depends on $\mathrm{sgn}(\nu)$. Setting $j' = j - \mathrm{sgn}(\nu)$ $(= j \pm 1)$, we take

$$\theta_j = \frac{U_{j'+1} - U_{j'}}{U_{j+1} - U_j}. \tag{16.33}$$

16.2.1 Linear systems

The natural generalization to linear systems is obtained by diagonalizing the system and applying the flux-limiter method to each of the resulting scalar equations. We can reexpress this in terms of the full system as follows. Suppose $A = R\Lambda R^{-1}$ with $R = [r_1|r_2|\cdots|r_m]$ and let

$$\alpha_j = R^{-1}(U_{j+1} - U_j) \tag{16.34}$$

be the vector with components α_{pj} for $p = 1, 2, \ldots, m$. Then we have

$$U_{j+1} - U_j = \sum_{p=1}^{m} \alpha_{pj} r_p \tag{16.35}$$

where r_p is the pth eigenvector of A. Set

$$\nu_p = \lambda_p k/h \tag{16.36}$$

and

$$\theta_{pj} = \frac{\alpha_{pj'}}{\alpha_{pj}} \tag{16.37}$$

where $j' = j - \mathrm{sgn}(\nu_p)$. Recall that the upwind flux has the form (13.15),

$$F_L(U;j) = \frac{1}{2}A(U_j + U_{j+1}) - \frac{1}{2}|A|(U_{j+1} - U_j). \tag{16.38}$$

while the Lax-Wendroff flux is

$$F_H(U;j) = \frac{1}{2}A(U_j + U_{j+1}) - \frac{1}{2}\frac{k}{h}A^2(U_{j+1} - U_j). \tag{16.39}$$

The difference between these is

$$F_H(U;j) - F_L(U;j) = \frac{1}{2}\left(|A| - \frac{k}{h}A^2\right)(U_{j+1} - U_j)$$

$$= \frac{1}{2}\sum_{p=1}^{m}(\text{sgn}(\nu_p) - \nu_p)\lambda_p\alpha_{pj}r_p$$

and so the flux-limiter method has a flux of the form

$$F(U;j) = F_L(U;j) + \frac{1}{2}\sum_{p=1}^{m}\phi(\theta_{pj})(\text{sgn}(\nu_p) - \nu_p)\lambda_p\alpha_{pj}r_p. \tag{16.40}$$

For nonlinear systems of equations a similar form is possible, based on the extension to nonlinear scalar problems indicated above coupled with a linearization based on Roe's approximate Riemann solution. Again the details are omitted since the resulting method is similar to the slope-limiter method presented below.

16.3 Slope-limiter methods

The second approach we will study is more geometric in nature. The basic idea is to generalize Godunov's method by replacing the piecewise constant representation of the solution by some more accurate representation, say piecewise linear.

Recall that Godunov's method can be viewed as consisting of the following three steps (although it is not typically implemented this way):

ALGORITHM 16.1.

1. Given data $\{U_j^n\}$, construct a function $\tilde{u}^n(x, t_n)$. (Piecewise constant in Godunov's method.)

2. Solve the conservation law exactly with this data to obtain $\tilde{u}^n(x, t_{n+1})$.

3. Compute cell averages of the resulting solution to obtain U_j^{n+1}.

To generalize this procedure, we replace Step 1 by a more accurate reconstruction, taking for example the piecewise linear function

$$\tilde{u}^n(x, t_n) = U_j^n + \sigma_j^n(x - x_j) \quad \text{on the cell } [x_{j-1/2}, x_{j+1/2}]. \tag{16.41}$$

Here σ_j^n is a slope on the jth cell which is based on the data U^n. For a system of equations, $\sigma_j^n \in \mathbb{R}^m$ is a vector of slopes for each component of u. Note that taking $\sigma_j^n = 0$ for all j and n recovers Godunov's method.

The cell average of $\tilde{u}^n(x, t_n)$ from (16.41) over $[x_{j-1/2}, x_{j+1/2}]$ is equal to U_j^n for any choice of σ_j^n. Since Steps 2 and 3 are also conservative, the overall method is conservative for any choice of σ_j^n.

For nonlinear problems we will generally not be able to perform Step 2 exactly. The construction of the exact solution $\tilde{u}^n(x,t)$ based on solving Riemann problems no longer works when $\tilde{u}^n(x,t_n)$ is piecewise linear. However, it is possible to approximate the solution in a suitable way, as will be discussed below.

The most interesting question is how do we choose the slopes σ_j^n? We will see below that for the linear advection equation with $a > 0$, if we make the natural choice

$$\sigma_j^n = \frac{U_{j+1}^n - U_j^n}{h} \qquad (16.42)$$

and solve the advection equation exactly in Step 2, then the method reduces to the Lax-Wendroff method. This shows that it is possible to obtain second order accuracy by this approach.

The oscillations which arise with Lax-Wendroff can be interpreted geometrically as being caused by a poor choice of slopes, leading to a piecewise linear reconstruction $\tilde{u}^n(x,t_n)$ with much larger total variation than the given data U^n. See Figure 16.3a for an example. We can rectify this by applying a **slope limiter** to (16.42), which reduces the value of this slope near discontinuities or extreme points, and is typically designed to ensure that

$$TV(\tilde{u}^n(\cdot,t_n)) \leq TV(U^n). \qquad (16.43)$$

The reconstruction shown in Figure 16.3b, for example, has this property. Since Steps 2 and 3 of Algorithm 16.1 are TVD, imposing (16.43) in Step 1 results in a method that is TVD overall, proving the following result.

THEOREM 16.4. *If the condition (16.43) is satisfied in Step 1 of Algorithm 16.1, then the method is TVD for scalar conservation laws.*

Methods of this type were first introduced by van Leer in a series of papers[88] through [92] where he develops the MUSCL Scheme (standing for "Monotonic Upstream-centered Scheme for Conservation Laws"). A variety of similar methods have since been proposed, e.g., [9], [26].

The reconstruction of Step 1 can be replaced by more accurate approximations as well. One can attempt to obtain greater accuracy by using quadratics, as in the piecewise parabolic method (PPM) of Colella and Woodward[10] or even higher order reconstructions as in the ENO (essentially nonoscillatory) methods[29], [34]. (See Chapter 17.)

Again, we will begin by considering the linear advection equation, and then generalize to nonlinear equations. For the linear equation we can perform Step 2 of Algorithm 16.1 exactly and obtain formulas that are easily reinterpreted as flux-limiter methods. This shows the close connection between the two approaches and also gives a more geometric interpretation to the TVD constraints discussed above.

For the advection equation, the exact solution $\tilde{u}^n(x,t_{n+1})$ is simply

$$\tilde{u}^n(x,t_{n+1}) = \tilde{u}^n(x - ak, t_n) \qquad (16.44)$$

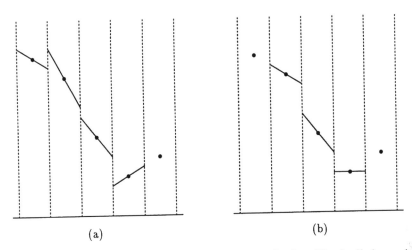

Figure 16.3. (a) Piecewise linear reconstruction using the Lax-Wendroff slopes (16.42). (b) Piecewise linear reconstruction using the minmod slopes (16.51).

and so computing the cell average in Step 3 of Algorithm 16.1 amounts to integrating the piecewise linear function defined by (16.41) over the interval $[x_{j-1/2} - ak, \; x_{j+1/2} - ak]$. It is straight forward to calculate that (for $a > 0$),

$$U_j^{n+1} = U_j^n - \nu(U_j^n - U_{j-1}^n) - \frac{1}{2}\nu(1-\nu)(h\sigma_j^n - h\sigma_{j-1}^n). \qquad (16.45)$$

If $\sigma_j^n \equiv 0$ this reduces to the upwind method, while for σ_j^n given by (16.42) it reduces to Lax-Wendroff, in the form (16.12).

Note that the numerical flux for (16.45) is

$$F(U;j) = aU_j + \frac{1}{2}a(1-\nu)h\sigma_j \qquad (16.46)$$

which has exactly the same form as the flux-limiter method (16.13) if we set

$$\sigma_j = \left(\frac{U_{j+1} - U_j}{h}\right)\phi_j. \qquad (16.47)$$

In this context the "flux-limiter" ϕ_j can be reinterpreted as a "slope-limiter".

More generally, for a of either sign we have

$$U_j^{n+1} = U_j^n - \nu(U_{j_1}^n - U_{j_1-1}^n) - \frac{1}{2}\nu(\mathrm{sgn}(\nu) - \nu)(h\sigma_{j_1} - h\sigma_{j_1-1}) \qquad (16.48)$$

where

$$j_1 = \begin{cases} j & \text{if } a > 0 \\ j+1 & \text{if } a < 0. \end{cases} \qquad (16.49)$$

The corresponding flux function is

$$F(U;j) = aU_{j_1} + \frac{1}{2}a(\text{sgn}(\nu) - \nu)h\sigma_{j_1}. \tag{16.50}$$

The first term here is simply the upwind flux and so again we have a direct correspondence between this formula and the flux-limiter formula (16.32).

EXERCISE 16.3. *Verify (16.45) and (16.48).*

In studying flux-limiter methods, we derived algebraic conditions that ϕ_j must satisfy to give a TVD method. Using the piecewise linear interpretation, we can derive similar conditions geometrically using the requirement (16.43).

One simple choice of slopes satisfying (16.43) is the so-called **minmod slope**,

$$\sigma_j = \frac{1}{h}\text{minmod}(U_{j+1} - U_j, \ U_j - U_{j-1}) \tag{16.51}$$

where the minmod function is defined by

$$\text{minmod}(a, b) = \begin{cases} a & \text{if } |a| < |b| \text{ and } ab > 0 \\ b & \text{if } |b| < |a| \text{ and } ab > 0 \\ 0 & \text{if } ab \leq 0 \end{cases} \tag{16.52}$$

$$= \frac{1}{2}(\text{sgn}(a) + \text{sgn}(b))\min(|a|, |b|).$$

Figure 16.3b shows the minmod slopes for one set of data.

We can rewrite the minmod slope-limiter method as a flux limiter method using (16.47) if we set

$$\phi(\theta) = \begin{cases} 0 & \text{if } \theta \leq 0 \\ \theta & \text{if } 0 \leq \theta \leq 1 \\ 1 & \text{if } \theta \geq 1. \end{cases} \tag{16.53}$$

$$= \max(0, \ \min(1, \theta))$$

Recall that $\theta_j = (U_j - U_{j-1})/(U_{j+1} - U_j)$ and so (16.47) with $\phi_j = \phi(\theta_j)$ and ϕ given by (16.53) reduces to (16.51). This limiter function ϕ lies along the lower boundary of Sweby's "second order TVD region" of Figure 16.1b.

Note that again $\phi(\theta) = 0$ for $\theta \leq 0$, which now corresponds to the fact that we set the slope σ_j to zero at extreme points of U, where the slopes $(U_{j+1} - U_j)/h$ and $(U_j - U_{j-1})/h$ have opposite signs. Geometrically, this is clearly required by (16.43) since any other choice will give a reconstruction $\tilde{u}^n(x, t_n)$ with total variation greater than $TV(U^n)$.

Although the minmod limiter (16.51) is a simple choice that clearly satisfies (16.43), it is more restrictive than necessary and somewhat larger slopes can often be taken without violating (16.43), and with greater resolution. Moreover, it is possible to violate (16.43) and still obtain a TVD method, since Step 3 of Algorithm 16.1 tends to reduce the total variation, and may eliminate overshoots caused in the previous steps.

A variety of other slope limiters have been developed. In particular, any of the flux limiters discussed above can be converted into slope limiters via (16.47). Conversely, a geometrically motivated slope limiter can often be converted into a flux limiter function $\phi(\theta)$. (In fact, van Leer's limiter (16.28) was initially introduced as a slope limiter in [89].)

16.3.1 Linear Systems

For a linear system of equations, we can diagonalize the system and apply the algorithm derived above to each decoupled scalar problem. Using the notation of Section 16.2.1, we let $V_j^n = R^{-1}U_j^n$ have components V_{pj} so that $U_j^n = \sum_{p=1}^m V_{pj}r_p$. We also set

$$j_p = \begin{cases} j & \text{if } \lambda_p > 0 \\ j+1 & \text{if } \lambda_p < 0 \end{cases} \tag{16.54}$$

generalizing j_1 defined in (16.49). Then the method (16.48) for each V_p takes the form

$$V_{pj}^{n+1} = V_{pj} - \nu_p(V_{pj_p} - V_{p,j_p-1}) + \frac{1}{2}\nu_p(\text{sgn}(\nu_p) - \nu_p)(h\beta_{pj_p} - h\beta_{p,j_p-1}) \tag{16.55}$$

where $\nu_p = k\lambda_p/h$ and β_{pj} is the slope for V_p in the jth cell. For example, we might take

$$\beta_{pj} = \frac{1}{h}\text{minmod}(V_{p,j+1} - V_{pj}, V_{pj} - V_{p,j-1}) = \frac{1}{h}\text{minmod}(\alpha_{pj}, \alpha_{p,j-1}). \tag{16.56}$$

Multiplying (16.55) by r_p and summing over p gives

$$
\begin{aligned}
U_j^{n+1} &= \sum_{p=1}^m V_{pj}^{n+1}r_p \\
&= U_j^n - \frac{k}{h}\left[\sum_{p=1}^m (V_{pj_p}\lambda_p r_p - V_{p,j_p-1}\lambda_p r_p)\right. \\
&\quad \left. + \frac{1}{2}\sum_{p=1}^m \lambda_p(\text{sgn}(\nu_p) - \nu_p)(h\sigma_{pj_p} - h\sigma_{p,j_p-1})\right]
\end{aligned} \tag{16.57}
$$

where σ_{pj} is the slope vector for the pth family,

$$\sigma_{pj} = \beta_{pj}r_p \in \mathbb{R}^m. \tag{16.58}$$

Recalling that the upwind flux $F_L(U;j)$ in (13.15) can also be written as

$$F_L(U;j) = \sum_{p=1}^m V_{pj_p}\lambda_p r_p, \tag{16.59}$$

we see that the flux for (16.57) has the form

$$F(U;j) = F_L(U;j) + \frac{1}{2} \sum_{p=1}^{m} \lambda_p(\text{sgn}(\nu_p) - \nu_p) h \sigma_{pj_p}. \qquad (16.60)$$

Notice that this is identical with the flux (16.40) for the flux-limiter method on a linear system if we identify

$$\sigma_{pj_p} = \phi(\theta_{pj}) \left(\frac{\alpha_{pj}}{h}\right) r_p, \qquad (16.61)$$

generalizing (16.47).

EXERCISE 16.4. *Verify that (16.61) holds when σ_{pj} is given by (16.56) and (16.58), θ_{pj} is given by (16.37), and ϕ is the minmod limiter (16.53).*

16.3.2   Nonlinear scalar equations.

In attempting to apply Algorithm 16.1 to a nonlinear problem, the principle difficulty is in Step 2, since we typically cannot compute the exact solution to the nonlinear equation with piecewise constant initial data. However, there are various ways to obtain approximate solutions which are sufficiently accurate that second order accuracy can be maintained.

I will describe one such approach based on approximating the nonlinear flux function by a linear function in the neighborhood of each cell interface, and solving the resulting linear equation exactly with the piecewise constant data. This type of approximation has already been introduced in the discussion of Roe's approximate Riemann solver in Chapter 14. The use of this approximation in the context of high resolution slope limiter methods for nonlinear scalar problems is studied in [26].

Here I will present the main idea in a simplified form, under the assumption that the data is monotone (say nonincreasing) and that $f'(u)$ does not change sign over the range of the data (say $f'(U_j^n) > 0$). A similar approach can be used near extreme points of U^n and sonic points, but more care is required and the formulas are more complicated (see [26] for details). Moreover, we will impose the time step restriction

$$\frac{k}{h} \max |f'(U_j^n)| < \frac{1}{2} \qquad (16.62)$$

although this can also be relaxed to the usual CFL limit of 1 with some modification of the method.

With the above assumptions on the data, we can define a piecewise linear function $\hat{f}(u)$ by interpolation the values $(U_j^n, f(U_j^n))$, as in Figure 16.4. We now define $\hat{u}^n(x,t)$ by solving the conservation law

$$\hat{u}_t + \hat{f}(\hat{u})_x = 0 \qquad (16.63)$$

for $t_n \leq t \leq t_{n+1}$, with the piecewise linear data (16.41). The evolution of $\hat{u}^n(x,t)$ is indicated in Figure 16.5.

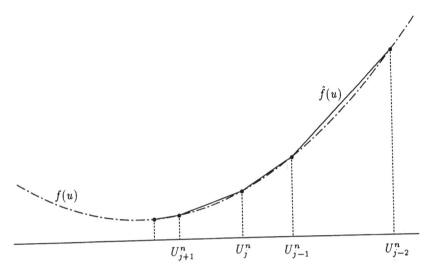

Figure 16.4. $\hat{f}(u)$ is a piecewise linear approximation of the flux function obtained by interpolation at the grid values.

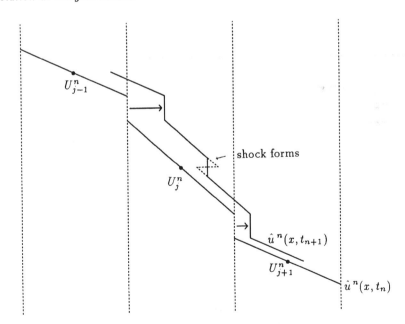

Figure 16.5. Evolution of the piecewise linear initial data with the conservation law with piecewise linear flux function $\hat{f}(\hat{u})$.

This flux is still nonlinear, but the nonlinearity has been concentrated at the points U_j^n. Shocks form immediately at the points x_j, but because of the time step restriction (16.62), these shocks do not reach the cell boundary during the time step. Hence we can easily compute the numerical flux

$$F(U^n; j) = \frac{1}{k} \int_{t_n}^{t_{n+1}} \hat{f}(\hat{u}^n(x_{j+1/2}, t)) \, dt. \tag{16.64}$$

At the cell boundary $x_{j+1/2}$, the solution values lie between U_j^n and U_{j+1}^n for $t_n \le t \le t_{n+1}$ and hence

$$\hat{f}(\hat{u}^n(x_{j+1/2}, t)) = f(U_j^n) + (\hat{u}^n(x_{j+1/2}, t) - U_j^n)\hat{a}_j, \tag{16.65}$$

where

$$\hat{a}_j = \frac{f(U_{j+1}) - f(U_j)}{U_{j+1} - U_j}. \tag{16.66}$$

The conservation law $\hat{u}_t + \hat{f}(\hat{u})_x = 0$ reduces to the advection equation $\hat{u}_t + \hat{a}_j \hat{u}_x = 0$ near $x_{j+1/2}$ and so

$$
\begin{aligned}
\hat{u}^n(x_{j+1/2}, t) &= \hat{u}^n(x_{j+1/2} - (t - t_n)\hat{a}_j, t_n) \\
&= U_j^n + \left(\frac{h}{2} - (t - t_n)\hat{a}_j \right) \sigma_j^n.
\end{aligned}
$$

Finally, again using (16.65), we compute the numerical flux (16.64) to be

$$
\begin{aligned}
F(U^n; j) &= \frac{1}{k} \int_{t_n}^{t_{n+1}} f(U_j^n) + \left(\frac{h}{2} - (t - t_n)\hat{a}_j \right) \sigma_j^n \, dt \\
&= f(U_j^n) + \frac{1}{2}\hat{a}_j \left(1 - \frac{k}{h}\hat{a}_j \right) h\sigma_j^n. \tag{16.67}
\end{aligned}
$$

For the linear advection equation this reduces to (16.46). For $\sigma_j^n \equiv 0$ this reduces to the upwind flux $f(U_j^n)$. With the choice of slopes (16.42) it reduces to

$$F(U^n; j) = \frac{1}{2}(f(U_j^n) + f(U_{j+1}^n)) - \frac{k}{2h}\left(\frac{(f(U_{j+1}^n) - f(U_j^n))^2}{U_{j+1}^n - U_j^n} \right), \tag{16.68}$$

which is a form of the Lax-Wendroff method for scalar nonlinear problems. Also notice the similarity of (16.67) to the flux-limiter formula (16.14). With the correspondence (16.47), (16.67) is clearly a generalization of (16.14) to the nonlinear case. (Note that $k\hat{a}_j/h$ is precisely ν_j from (16.29).)

To obtain a high resolution TVD method of this form, we can again choose the slope σ_j as in the linear case, for example using the minmod slope (16.51), so that the total variation bound (16.43) is satisfied.

Notice that although we do not solve our original conservation law exactly in Step 2 of Algorithm 16.1, we do obtain $\hat{u}^n$ as the exact solution to a modified conservation law, and hence $\hat{u}^n$ is total variation diminishing. By using a slope limiter that enforces (16.43), we obtain an overall method for the nonlinear scalar problem that is TVD.

16.3.3   Nonlinear Systems

The natural way to generalize this method to a nonlinear system of equations is to linearize the equations in the neighborhood of each cell interface $x_{j+1/2}$ and apply the method of Section 16.3.1 to some linearized system

$$u_t + \hat{A}_j u_x = 0. \tag{16.69}$$

This is what we did in the scalar case, when the linearization was given by (16.66). We have already seen how to generalize (16.66) to a system of equations in our discussion of Roe's approximate Riemann solution (Section 14.2). We take $\hat{A}_j = \hat{A}(U_j^n, U_{j+1}^n)$, where $\hat{A}$ is some Roe matrix satisfying conditions (14.19). We denote the eigenvalues and eigenvectors of $\hat{A}_j$ by $\hat{\lambda}_{pj}$ and $\hat{r}_{pj}$ respectively, so that

$$\hat{A}_j \hat{r}_{pj} = \hat{\lambda}_{pj} \hat{r}_{pj} \qquad \text{for } p = 1, 2, , \ldots, m.$$

Recall that the flux function for Godunov's method with Roe's approximate Riemann solver is given by (14.22), which we rewrite as

$$F_L(U; j) = f(u_j) + \sum_{p=1}^{m} \hat{\lambda}_{pj}^- \alpha_{pj} \hat{r}_{pj} \tag{16.70}$$

where $\hat{\lambda}_{pj}^- = \min(\hat{\lambda}_p, 0)$ and α_{pj} is the coefiicient of $\hat{r}_{pj}$ in an eigenvector expansion of $U_{j+1} - U_j$,

$$U_{j+1} - U_j = \sum_{p=1}^{m} \alpha_{pj} \hat{r}_{pj}. \tag{16.71}$$

If we set $\nu_{pj} = k \hat{\lambda}_{pj}/h$, then the natural generalization of (16.60) to the nonlinear system is

$$F(U; j) = F_L(U; j) + \frac{1}{2} \sum_{p=1}^{m} \hat{\lambda}_{pj} (\text{sgn}(\nu_{pj}) - \nu_{pj}) h \sigma_{pj_p} \tag{16.72}$$

where $\sigma_{pj} \in \mathbb{R}^m$ is some slope vector for the pth family. Note that combining (16.56) and (16.58) gives the following form for σ_{pj} in the case of a linear system:

$$\sigma_{pj} = \frac{1}{h} \text{minmod}(\alpha_{pj}, \alpha_{p,j-1}) r_p. \tag{16.73}$$

For our nonlinear method, r_p is replaced by $\hat{r}_{pj}$, which now varies with j. Moreover, it is only the vectors $\alpha_{pj} \hat{r}_{pj}$ and $\alpha_{p,j-1} \hat{r}_{p,j-1}$ that are actually computed in Roe's method, not the normalized $\hat{r}$ and coefficient α separately, and so the natural generalization of (16.73) to the nonlinear case is given by

$$\sigma_{pj} = \frac{1}{h} \text{minmod}(\alpha_{pj} \hat{r}_{pj}, \alpha_{p,j-1} \hat{r}_{p,j-1}) \tag{16.74}$$

where the minmod function is not applied componentwise to the vector arguments.

Of course the minmod function in (16.74) could be replaced by any other slope limiter, again applied componentwise. The "high resolution" results presented in Figure 1.4 were computed by this method with the superbee limiter.

In deriving this method we have ignored the entropy condition. Since we use Roe's approximate Riemann solution, which replaces rarefaction waves by discontinuities, we must in practice apply an entropy fix as described in Section 14.2.2. The details will not be presented here.

The flux (16.72) with slope (16.74) gives just one high resolution method for nonlinear systems of conservation laws. It is not the most sophisticated or best, but it is a reasonable method and our development of it has illustrated many of the basic ideas used in many other methods. The reader is encouraged to explore the wide variety of methods available in the literature.

17 Semi-discrete Methods

The methods discussed so far have all been fully discrete methods, discretized in both space and time. At times it is useful to consider the discretization process in two stages, first discretizing only in space, leaving the problem continuous in time. This leads to a system of ordinary differential equations in time, called the "semi-discrete equations". We then discretize in time using any standard numerical method for systems of ordinary differential equations. This approach of reducing a PDE to a system of ODEs, to which we then apply an ODE solver, is often called the **method of lines**.

This approach is particularly useful in developing methods with order of accuracy greater than 2, since it allows us to decouple the issues of spatial and temporal accuracy. We can define high order approximations to the flux at a cell boundary at one instant in time using high order interpolation in space, and then achieve high order temporal accuracy by applying any of the wide variety of high order ODE solvers.

This approach is also useful in extending methods to two or more space dimensions, as we will see in Chapter 18.

17.1 Evolution equations for the cell averages

Let $U_j(t)$ represent a discrete approximation to the cell average of u over the jth cell at time t, i.e.,

$$U_j(t) \approx \bar{u}_j(t) \equiv \frac{1}{h} \int_{x_{j-1/2}}^{x_{j+1/2}} u(x,t) \, dx. \tag{17.1}$$

We know from the integral form of the conservation law (2.15) that the cell average $\bar{u}_j(t)$ evolves according to

$$\bar{u}_j'(t) = -\frac{1}{h} \left[f(u(x_{j+1/2}, t)) - f(u(x_{j-1/2}, t)) \right] \equiv H(u(\cdot, t); j). \tag{17.2}$$

We now let $\mathcal{F}(U(t); j)$ represent an approximation to $f(u(x_{j+1/2}, t))$, obtained from the discrete data $U(t) = \{U_j(t)\}$. For example, we might solve the Riemann problem with data $U_j(t)$ and $U_{j+1}(t)$ for the intermediate state $u^*(U_j(t), U_{j+1}(t))$ and then set

$$\mathcal{F}(U(t); j) = f\left(u^*(U_j(t), U_{j+1}(t))\right). \tag{17.3}$$

Replacing the true fluxes in (17.2) by $\mathcal{F}$ and the cell average $\bar{u}_j(t)$ by $U_j(t)$, we obtain a discrete system of ordinary differential equations for the $U_j(t)$,

$$U_j'(t) = -\frac{1}{h}[\mathcal{F}(U(t); j) - \mathcal{F}(U(t); j-1)] \equiv \mathcal{H}(U(t); j). \tag{17.4}$$

Note that this is a coupled system of equations since each of the fluxes $\mathcal{F}$ depends on two or more of the $U_i(t)$.

We can now discretize in time. For example, if we discretize (17.4) using Euler's method with time step k, and let U_j^n now represent our fully-discrete approximation to $U_j(t_n)$, then we obtain

$$\begin{aligned} U_j^{n+1} &= U_j^n + k\mathcal{H}(U(t); j) \tag{17.5} \\ &= U_j^n - \frac{k}{h}[\mathcal{F}(U^n; j) - \mathcal{F}(U^n; j-1)], \end{aligned}$$

which is in the familiar form of a conservative method. In particular, if $\mathcal{F}$ is given by (17.3), then (17.5) is simply Godunov's method. More generally, however, $\mathcal{F}(U(t); j)$ represents an approximation to the value of $f(u(x_{j+1/2}, t))$ at one point in time, whereas the standard numerical flux $F(U; j)$ used before has always represented an approximation to the average of $f(u(x_{j+1/2}, t))$ over the time interval $[t_n, t_{n+1}]$. In Godunov's method these are the same, since $u(x_{j+1/2}, t)$ is constant in time in the Riemann solution.

To obtain higher order accuracy, we must make two improvements: the value $\mathcal{F}$ obtained by piecewise constant approximations in (17.3) must be improved, and the first order accurate Euler method must be replaced by a higher order time-stepping method. One advantage of the method of lines approach is that the spatial and temporal accuracy are decoupled, and can be considered separately. This is particularly useful in several space dimensions.

One way to obtain greater spatial accuaracy is to use a piecewise linear approximation in defining $\mathcal{F}$. From the data $\{U_j(t)\}$ we can construct a piecewise linear function $\tilde{u}(x)$ using slope limiters as discussed in Chapter 16. Then at the interface $x_{j+1/2}$ we have values on the left and right from the two linear approximations in each of the neighboring cells (see Figure 17.1). Denote these values by

$$U_{j+1/2}^L = U_j + \frac{h}{2}\sigma_j \tag{17.6}$$

$$U_{j+1/2}^R = U_{j+1} - \frac{h}{2}\sigma_{j+1} \tag{17.7}$$

A second order accurate approximation to the flux at this cell boundary at time t is then obtained by solving the Riemann problem with left and right states given by these two values, and setting

$$\mathcal{F}(U; j) = f\left(u^*(U_{j+1/2}^L, U_{j+1/2}^R)\right). \tag{17.8}$$

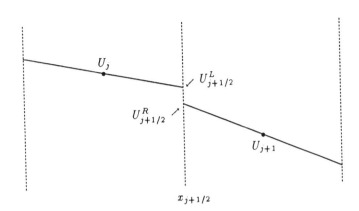

$x_{j+1/2}$

Figure 17.1. Piecewise linear reconstruction of $\tilde{u}(x)$ *used to define left and right states at* $x_{j+1/2}$.

This type of semi-discrete MUSCL scheme is discussed in more detail by Osher[59].

If we use this flux in the fully discrete method (17.5), then the method is second order accurate in space but only first order accurate in time, i.e., the global error is $O(h^2 + k)$, since the time discretization is still Euler's method. For time-dependent problems this improvement in spatial accuracy alone is usually not advantageous, but for steady state problems this type of method will converge as time evolves to a second order accurate approximation to the steady state solution, in spite of the fact that it is not second order accurate in time along the way.

To obtain a method that is second order accurate in time as well as space, we can discretize the ODEs (17.4) using a second order accurate ODE method. One possibility is the 2-stage Runge-Kutta method

$$U_j^* = U_j^n - \frac{k}{2h}[\mathcal{F}(U^n;j) - \mathcal{F}(U^n;j-1)] \tag{17.9}$$

$$U_j^{n+1} = U_j^n - \frac{k}{h}[\mathcal{F}(U^*;j) - \mathcal{F}(U^*;j-1)]. \tag{17.10}$$

Note that this requires solving two Riemann problems at each cell boundary in each time step. For further discussion of time discretizations, see for example [39], or [72].

17.2 Spatial accuracy

We first analyze the spatial accuracy of the semi-discrete method (17.4), i.e., we compute the error $U_j(t) - \bar{u}_j(t)$ in the *exact* solution of the system of ODEs. Let $e_j(t)$ represent the global error,

$$e_j(t) = U_j(t) - \bar{u}_j(t). \tag{17.11}$$

Then subtracting (17.2) from (17.4) gives an evolution equation for $e_j(t)$:

$$e_j'(t) = \mathcal{H}(U(t); j) - H(u(\cdot, t); j). \tag{17.12}$$

If we let $\bar{u}(t)$ be the vector with components $\bar{u}_j(t)$, then we can rewrite (17.12) as

$$e_j'(t) = (\mathcal{H}(U(t); j) - \mathcal{H}(\bar{u}(t); j)) + (\mathcal{H}(\bar{u}(t); j) - H(u(\cdot, t); j)). \tag{17.13}$$

We define the local truncation error $L_h(j, t)$ by

$$L_h(j, t) = \mathcal{H}(\bar{u}(t); j) - H(u(\cdot, t); j) \tag{17.14}$$

for smooth functions $u(x, t)$ satisfying the ODEs (17.2). This is consistent with previous definitions of local truncation error: it is the difference between the discrete operator applied to the true solution and the exact operator applied to the true solution. It can be shown that if $\mathcal{H}(U; j)$ is Lipschitz continuous and pth order accurate in the sense that

$$|L_h(j, t)| \le Ch^p, \tag{17.15}$$

then the semi-discrete method (17.4) is pth order accurate,

$$|e_j(t)| = O(h^p) \qquad \text{as } h \to 0. \tag{17.16}$$

17.3   Reconstruction by primitive functions

To obtain high spatial accuracy we need to define $\mathcal{F}$ in such a way that $\mathcal{F}(\bar{u}(t); j)$ is a good approximation to $f(u(x_{j+1/2}, t))$. Recall that $\bar{u}(t)$ is the vector of exact *cell averages*, and from these we want to obtain a value $U_{j+1/2}$ that approximates the *pointwise* value $u(x_{j+1/2}, t)$. One approach to this was outlined above: define $U^L_{j+1/2}$ and $U^R_{j+1/2}$ using slope-limited piecewise linears and then set

$$U_{j+1/2} = u^*(U^L_{j+1/2}, U^R_{j+1/2}).$$

To obtain higher order accuracy we can take the same approach but define $U^{L,R}_{j+1/2}$ via some higher order polynomial approximation to u over the cells to the left and right of $x_{j+1/2}$.

This raises the following question: *Given only the cell averages $\bar{u}_j(t)$, how can we construct a polynomial approximation to u that is accurate pointwise to high order?*

A very elegant solution to this problem uses the **primitive function** for $u(x, t)$. This approach was apparently first introduced by Colella and Woodward[10] in their PPM method and has since been used in a variety of other methods, particularly the ENO methods discussed below.

At a fixed time t, the primitive function $w(x)$ is defined by

$$w(x) = \int_{x_{1/2}}^{x} u(\xi, t)\, d\xi. \tag{17.17}$$

The lower limit $x_{1/2}$ is arbitrary, any fixed point could be used. Changing the lower limit only shifts $w(x)$ by a constant, and the property of w that we will ultimately use is that

$$u(x, t) = w'(x), \tag{17.18}$$

which is unaffected by a constant shift. Equation (17.18) allows us to obtain pointwise values of u if we have a good approximation to w.

Now the crucial observation is that knowing *cell averages* of u gives us *pointwise* values of w at the particular points $x_{j+1/2}$. Set

$$W_j = w(x_{j+1/2}) = \int_{x_{1/2}}^{x_{j+1/2}} u(\xi, t)\, d\xi. \tag{17.19}$$

This is h times the average of u over a collection of j cells, and hence

$$W_j = h \sum_{i=1}^{j} \bar{u}_i(t).$$

Of course this only gives us pointwise values of w at the points $x_{j+1/2}$, but it gives us the *exact* values at these points (assuming we start with the exact cell averages $\bar{u}_j$, as we do in computing the truncation error). If w is sufficiently smooth (i.e., if u is sufficiently smooth), we can then approximate w more globally to arbitrary accuracy using polynomial interpolation. In particular, to approximate w in the jth cell $[x_{j-1/2}, x_{j+1/2}]$, we can use an interpolating polynomial of degree q passing through some $q+1$ points $W_{j-i}, W_{j-i+1}, \ldots, W_{j-i+q}$ for some i. (The choice of i is discussed below.) If we call this polynomial $p_j(x)$, then we have

$$p_j(x) = w(x) + O(h^{q+1}) \tag{17.20}$$

for $x \in [x_{j-1/2}, x_{j+1/2}]$, provided $w \in C^{q+1}$ (which requires $u(\cdot, t) \in C^q$).

Using the relation (17.18), we can obtain an approximation to $u(x, t)$ by differentiating $p_j(x)$. We lose one order of accuracy by differentiating the interpolating polynomial, and so

$$p_j'(x) = u(x, t) + O(h^q) \qquad \text{on } [x_{j-1/2}, x_{j+1/2}].$$

We can now use this to obtain approximations to u at the left and right cell interfaces, setting

$$\begin{aligned} U_{j-1/2}^{R} &= p_j'(x_{j-1/2}) \\ U_{j+1/2}^{L} &= p_j'(x_{j+1/2}). \end{aligned}$$

Performing a similar reconstruction on the cell $[x_{j+1/2}, x_{j+3/2}]$ gives $p_{j+1}(x)$ and we set

$$U^R_{j+1/2} = p'_{j+1}(x_{j+1/2})$$

and then define $\mathcal{F}$ as in (17.8). This gives spatial accuracy of order q for sufficiently smooth u.

17.4   ENO schemes

In the above description of the interpolation process, the value of i was left unspecified (recall we interpolate $W_{j-i}, \ldots, W_{j-i+q}$ to approximate u on $[x_{j-1/2}, x_{j+1/2}]$). When $u(\cdot, t) \in C^q$ in the vicinity of x_j, interpolation based on any value of i between 1 and q will give qth order accuracy. However, for a high resolution method we must be able to automatically cope with the possibility that the data is not smooth. Near discontinuities we do not expect to maintain the high order of accuracy, but want to choose a stencil of interpolation points that avoids introducing oscillations. It is well known that a high degree polynomial interpolant can be highly oscillatory even on smooth data, and certainly will be on nonsmooth data.

In the piecewise linear version described initially, this was accomplished by using a slope-limiter, for example the minmod slope which compares linear interpolants based on cells to the left and right and takes the one that is less steep. This gives a global piecewise linear approximation that is nonoscillatory in the sense that its total variation is no greater than that of the discrete data.

This same idea can be extended to higher order polynomials by choosing the value of i for each j so that the interpolant through $W_{j-i}, \ldots, W_{j-i+q}$ has the least oscillation over all possible choices $i = 1, \ldots, q$. This is the main idea in the ENO (essentially nonoscillatory) methods developed by Chakravarthy, Engquist, Harten and Osher, and complete details, along with several variations, can be found in their papers[28], [29], [35], [74], [73].

One variant uses the following procedure. Start with the linear function passing through W_{j-1} and W_j to define $p^{(1)}_j(x)$ (where superscripts now indicate the degree of the polynomial). Next compute the divided difference based on $\{W_{j-2}, W_{j-1}, W_j\}$ and the divided difference based on $\{W_{j-1}, W_j, W_{j+1}\}$. Either of these can be used to extend $p^{(1)}_j(x)$ to a quadratic polynomial using the Newton form of the interpolating polynomial. We define $p^{(2)}_j(x)$ by choosing the divided difference that is smaller in magnitude.

We continue recursively in this manner, adding either the next point to the left or to the right to our stencil depending on the magnitude of the divided difference, until we have a polynomial of degree q based on some $q + 1$ points.

Note that the first order divided differences of W are simply the values $\bar{u}_j$,

$$\frac{W_j - W_{j-1}}{h} = \bar{u}_j,$$

and so divided differences of W are directly related to divided differences of the cell averages $\bar{u}_j$. In practice we need never compute the W_j. (The zero order divided difference, W_j itself, enters $p_j(x)$ only as the constant term which drops out when we compute $p'_j(x)$.)

EXERCISE 17.1. *Show that the ENO method described above with $q = 2$ (quadratic interpolation of W) gives a piecewise linear reconstruction of u with slopes that agree with the minmod formula (16.51).*

18 Multidimensional Problems

Most practical problems are in two or more space dimensions. So far, we have only considered the one-dimensional (1D) problem. To some extent the 1D methods and theory can be applied to problems in more than one space dimension, and some of these extensions will be briefly described in this chapter. We look at the two-dimensional (2D) case to keep the notation simple, but the same ideas can be used in three dimensions as well.

In two space dimensions a system of conservation laws takes the form

$$u_t + f(u)_x + g(u)_y = 0 \tag{18.1}$$

where $u = u(x, y, t) \in \mathbb{R}^m$. Typically the problem geometry is complicated — we may want to calculate the flow over an airfoil or other body, or through a nozzle or turbine. In this case simply defining an appropriate numerical grid on which to compute a solution can be a difficult problem, and the proper application of the physically relevant boundary conditions introduces additional difficulties. The computational fluid dynamics books referenced in Chapter 1 discuss these problems in some detail. Here I will only discuss the case in which a rectangular, or Cartesian, grid is used, meaning that the grid points are of the form (x_i, y_j) as i, j range through the integers, with

$$x_i = ih, \qquad y_j = jh. \tag{18.2}$$

For convenience I assume the grid spacing is the same in both coordinate directions, but this is not essential.

For conservation laws, it is again better to consider grid cells rather than grid points, and let $\bar{u}_{ij}$ be the cell average of $u(x, y, t)$ over the (i, j) cell $[x_{i-1/2}, x_{i+1/2}] \times [y_{j-1/2}, y_{j+1/2}]$. Then the integral form of the conservation law takes the form

$$\frac{d}{dt} \bar{u}_{ij}(t) = -\frac{1}{h^2} \int_{y_{j-1/2}}^{y_{j+1/2}} f(u(x_{i+1/2}, y, t)) - f(u(x_{i-1/2}, y, t))\, dy$$

$$-\frac{1}{h^2} \int_{x_{i-1/2}}^{x_{i+1/2}} g(u(x, y_{j+1/2}, t)) - g(u(x, y_{j-1/2}, t))\, dx. \tag{18.3}$$

Numerical methods based on this viewpoint (updating cell averages based on fluxes through the boundary of the cell) are often referred to as **finite volume methods**.

18.1 Semi-discrete methods

One approach to solving the 2D problem is to first introduce a semi-discrete approximation, as we did above in 1D, and then discretize the resulting system of ODEs. We define pointwise fluxes $\mathcal{F}$ and $\mathcal{G}$ approximating the appropriate integrals above:

$$\mathcal{F}(U;i,j) \approx \frac{1}{h}\int_{y_{j-1/2}}^{y_{j+1/2}} f(u(x_{i+1/2},y,t))\,dy$$

$$\mathcal{G}(U;i,j) \approx \frac{1}{h}\int_{x_{j-1/2}}^{x_{j+1/2}} f(u(x,y_{j+1/2},t))\,dx$$

and introduce approximations $U_{ij}(t)$ to $\bar{u}_{ij}(t)$. We then obtain the ODEs

$$U'_{ij}(t) = -\frac{1}{h}[\mathcal{F}(U;i,j) - \mathcal{F}(U;i-1,j) + \mathcal{G}(U;i,j) - \mathcal{G}(U;i,j-1)] \qquad (18.4)$$

as our semi-discrete method.

The simplest method of this form, a generalization of Godunov's method to 2D, is obtained by setting $\mathcal{F}(U;i,j)$ equal to $f(u^*)$, where u^* is the intermediate state obtained in solving the one dimensional Riemann problem

$$u_t + f(u)_x = 0 \qquad (18.5)$$

with data $U_{ij}(t)$ and $U_{i+1,j}(t)$. Similarly, $\mathcal{G}(U;i,j)$ is equal to $g(u^*)$, where u^* is now obtained by solving the Riemann problem

$$u_t + g(u)_y = 0 \qquad (18.6)$$

with data $U_{ij}(t)$ and $U_{i,j+1}(t)$. Discretizing in time using Euler's method then gives a 2D generalization of Godunov's method, which is first order accurate.

Higher order accuracy can be obtained by introducing slopes in the x and y directions (with slope limiters for nonoscillatory results) in order to calculate more accurate values for $\mathcal{F}$ and $\mathcal{G}$. Euler's method must then be replaced by a more accurate time-stepping method in order to obtain second order accuracy in time as well as space.

EXERCISE 18.1. *Show that the two-dimensional version of Godunov's method described above has a stability limit no greater than*

$$\frac{k}{h}\max(|f'(u)|, |g'(u)|) \leq \frac{1}{2}, \qquad (18.7)$$

by considering the linear problem

$$u_t + u_x + u_y = 0$$

with data

$$U_j^0 = \begin{cases} 1 & i+j < 0, \\ 0 & i+j \geq 0. \end{cases}$$

18.2   Splitting methods

Another approach to developing numerical methods in two space dimensions is to use any of the fully discrete one-dimensional methods developed in the preceeding chapters, and to apply them alternately on one-dimensional problems in the x and y directions.

The possibility that this might be successful is best illustrated by considering the 2D scalar advection equation

$$u_t + au_x + bu_y = 0 \tag{18.8}$$

with initial data

$$u(x, y, 0) = u_0(x, y).$$

The exact solution is

$$u(x, y, t) = u_0(x - at, y - bt).$$

Suppose we attempt to solve this equation by instead solving a pair of 1D problems. We first solve

$$u_t^* + au_x^* = 0 \tag{18.9}$$

with data

$$u^*(x, y, 0) = u_0(x, y)$$

to obtain $u^*(x, y, t)$. We then use this function of x and y as initial data and solve the 1D problem

$$u_t^{**} + bu_y^{**} = 0 \tag{18.10}$$

with data

$$u^{**}(x, y, 0) = u^*(x, y, t)$$

to obtain $u^{**}(x, y, t)$.

How well does $u^{**}(x, y, t)$ approximate $u(x, y, t)$? We easily compute that

$$u^*(x, y, t) = u_0(x - at, y)$$

is the exact solution of the advection equation (18.9), and hence solving (18.10) gives

$$
\begin{aligned}
u^{**}(x, y, t) &= u^{**}(x, y - bt, 0) \\
&= u^*(x, y - bt, t) \\
&= u_0(x - at, y - bt) \\
&= u(x, y, t).
\end{aligned}
$$

We see that u^{**} is in fact the exact solution, in this linear scalar case.

To develop a more general theory, let S_t^x represent the exact solution operator for the 1D equation (18.9) over time t, so that we can write

$$
u^*(\cdot, \cdot, t) = S_t^x u^*(\cdot, \cdot, 0).
$$

Similarly, let S_t^y be the solution operator for the advection equation (18.10) and let S_t be the solution operator for the full problem (18.8). Then what we have found above is that

$$
\begin{aligned}
u(\cdot, \cdot, t) &= S_t u_0 \\
&= S_t^y S_t^x u_0,
\end{aligned}
$$

i.e.,

$$
S_t = S_t^y S_t^x \tag{18.11}
$$

for all times t. The solution operator for the 2D problem splits as the product of two 1D solution operators.

For the advection equation we can write the solution operator formally as

$$
S_t^x = e^{-ta\partial_x}, \tag{18.12}
$$

where ∂_x is short for $\frac{\partial}{\partial x}$ and the operator on the right hand side is defined by the power series

$$
e^{-ta\partial_x} = 1 - ta\partial_x + \frac{1}{2}t^2 a^2 \partial_x^2 - \cdots.
$$

The relation (18.12) follows from the Taylor series expansion of the solution $u^*(x, y, t)$ to (18.9),

$$
\begin{aligned}
u^*(x, y, t) &= u^*(x, y, 0) + t\partial_t u^*(x, y, 0) + \frac{1}{2}t^2 \partial_t^2 u^*(x, y, 0) + \cdots \\
&= \left(1 + t\partial_t + \frac{1}{2}t^2 \partial_t^2 + \cdots\right) u^*(x, y, 0)
\end{aligned}
$$

and the fact that $u_t^* = -au_x^*$, which implies more generally that $\partial_t^q u^* = (-a)^q \partial_x^q u^*$. Similarly,

$$
S_t^y = e^{-tb\partial_y}
$$

and

$$S_t = e^{-t(a\partial_x + b\partial_y)}.$$

The relation (18.11) then follows from the product rule for exponentials.

In order to use this splitting in a numerical method, we simply replace the exact solution operators S_t^x and S_t^y by numerical methods $\mathcal{H}_k^x$ and $\mathcal{H}_k^y$ for the corresponding 1D problems over each time step of length k. This gives a numerical method of the form

$$U^{n+1} = \mathcal{H}_k^y \mathcal{H}_k^x U^n. \tag{18.13}$$

If $\mathcal{H}_k^x$ and $\mathcal{H}_k^y$ both represent pth order accurate methods, then the split method (18.13) will also be pth order accurate for this scalar linear problem, since there is no splitting error introduced.

Now consider a linear system of equations in 2D,

$$u_t + Au_x + Bu_y = 0, \tag{18.14}$$

and define S_t^x, S_t^y and S_t as before, so

$$\begin{aligned} S_t^x &= e^{-tA\partial_x} \\ S_t^y &= e^{-tB\partial_y} \\ S_t &= e^{-t(A\partial_x + B\partial_y)}. \end{aligned}$$

For matrices, the product rule no longer holds exactly, except in the case where A and B commute. Expansion of these operators in power series and formal multiplication of the series shows that in general

$$e^{-tB\partial_y} e^{-tA\partial_x} = e^{-t(A\partial_x + B\partial_y)} - \frac{1}{2}t^2(AB - BA)\partial_x\partial_y + O(t^3).$$

If we replace the 2D solution operator by the product of 1D solution operators, we introduce an error which depends on the commutator $AB - BA$. However, this error is $O(t^2)$ for small t, which suggests that using a splitting of this form in each time step of length k will introduce a splitting error of magnitude $O(k^2)$. After dividing by k this translates into a truncation error which is $O(k)$, leading to a method that is consistent and first order accurate for the original 2D equation (18.14).

Strang[81] pointed out that the accuracy of splitting methods can be increased to second order if we approximate the 2D solution operator by a slightly different product of 1D solution operators. The **Strang splitting** has the form

$$e^{-\frac{1}{2}kA\partial_x} e^{-kB\partial_y} e^{-\frac{1}{2}kA\partial_x} = e^{-k(A\partial_x + B\partial_y)} + O(k^3). \tag{18.15}$$

In a numerical method based on this splitting, the exact 1D solution operators $e^{-\frac{1}{2}kA\partial_x}$ and $e^{-kB\partial_y}$ are replaced by 1D numerical methods, giving a method with the form

$$U^{n+1} = \mathcal{H}_{k/2}^x \mathcal{H}_k^y \mathcal{H}_{k/2}^x U^n. \tag{18.16}$$

If each of the methods $\mathcal{H}^x$ and $\mathcal{H}^y$ is second order accurate, then so is the split method (18.16).

EXERCISE 18.2. *Compute the local truncation error for the method (18.16) in terms of the truncation errors of the 1D methods and the splitting error.*

At first glance the Strang split method (18.16) seems to be more expensive to implement than the first order method (18.13), since three applications of 1D operators are required in each time step rather than two. In practice, however, several time steps are taken together and the $\mathcal{H}^x_{k/2}$ operators can be combined to yield

$$U^n = \mathcal{H}^x_{k/2}(\mathcal{H}^y_k\mathcal{H}^x_k)^{n-1}\mathcal{H}^y_k\mathcal{H}^x_{k/2}U^0.$$

In this form the expense of this method is little more than that of the first order method

$$U^n = (\mathcal{H}^y_k\mathcal{H}^x_k)^n U^0.$$

It is only at the beginning and end of the computation (and at any intermediate times where output is desired) that the half step operators $\mathcal{H}^x_{k/2}$ must be employed.

For a nonlinear system of conservation laws, a fractional step method still takes the form (18.16), where $\mathcal{H}^x_k$ and $\mathcal{H}^y_k$ now represent numerical methods for the corresponding 1D conservation laws (18.5) and (18.6). If each of these methods is second order accurate, then the split method (18.16) remains second order accurate on smooth solutions of nonlinear problems. This can be shown by appropriate expansions in Taylor series[48].

For nonsmooth solutions, some results for scalar conservation laws have been obtained by Crandall and Majda[15]. In this case we let S^x_t, S^y_t and S_t represent the exact solution operators yielding the (unique) entropy-satisfying weak solutions. Then the 1D solution operators can be combined using either the Strang splitting or the simpler first order splitting to yield an approximation to the 2D solution operator which converges as $k \to 0$.

THEOREM 18.1 (CRANDALL AND MAJDA[15]). *If the exact solution operator is used in each step of the fractional step procedure, then the method converges to the weak solution of the 2D scalar conservation law, i.e.,*

$$\|S_T u_0 - [S^y_k S^x_k]^n u_0\|_1 \to 0$$

and

$$\left\|S_T u_0 - \left[S^x_{k/2} S^y_k S^x_{k/2}\right]^n u_0\right\|_1 \to 0$$

as $k \to 0$ and $n \to \infty$ with $nk = T$.

This shows that there is hope that numerical methods based on these splittings will also converge to the solution of the 2D problem. If we use monotone methods for each 1D problem, then this can in fact be shown:

THEOREM 18.2 (CRANDALL AND MAJDA[15]). *If the exact solution operators S^x, S^y in the above theorem are replaced by monotone methods $\mathcal{H}^x$, $\mathcal{H}^y$ for the 1D conservation laws, then the results still hold.*

18.3 TVD Methods

Recall that monotone methods are at most first order accurate. We would like to prove convergence of methods which have a higher order of accuracy. In one dimension we found it convenient to consider the total variation of the function. In two dimensions the true solution to a scalar conservation law is still total variation diminishing, where the total variation is now defined as in (15.6) but with t replaced by y:

$$TV(u) \; = \; \limsup_{\epsilon \to 0} \frac{1}{\epsilon} \int_{-\infty}^{\infty} \int_{-\infty}^{\infty} |u(x + \epsilon, y) - u(x, y)| \, dx \, dy$$

$$+ \limsup_{\epsilon \to 0} \frac{1}{\epsilon} \int_{-\infty}^{\infty} \int_{-\infty}^{\infty} |u(x, y + \epsilon) - u(x, y)| \, dx \, dy. \qquad (18.17)$$

We can define the total variation of a discrete grid function analogously by

$$TV(U) = h \sum_{i=-\infty}^{\infty} \sum_{j=-\infty}^{\infty} \left[|U_{i+1,j} - U_{ij}| + |U_{i,j+1} - U_{ij}| \right]. \qquad (18.18)$$

If the discrete method is TVD, then we again have all of our approximations lying in some compact set and we obtain a convergence result analogous to Theorem 15.2. In one dimension, it is possible to develop second order accurate TVD methods and so this is a very useful concept. Unfortunately, in two dimensions it turns out that this is not possible.

THEOREM 18.3 (GOODMAN AND LEVEQUE[25]). *Except in certain trivial cases, any method that is TVD in two space dimensions is at most first order accurate.*

In spite of this negative result, numerical methods obtained using high resolution 1D methods combined with the Strang splitting typically work very well in practice. They remain second order accurate on smooth solutions and usually give nonoscillatory and sharp solutions in spite of the fact that they are not technically TVD methods. As in the case of systems of equations in one space dimension, the mathematical theory is lagging behind the state-of-the-art computational methods.

18.4 Multidimensional approaches

The use of 1D methods via one of the approaches discussed above has various advantages: the methods are relatively easy to develop in two or three dimensions, and to some extent the 1D theory carries over. However, there are obvious disadvantages with this approach as well. Two or three dimensional effects do play a strong role in the behavior of the solution locally, and any approach that only solves 1D Riemann problems in the coordinate directions is clearly not using all the available information. We are also introducing a directional bias in the coordinate directions, and wave propagation is typically anisotropic as a result. Shock waves at a large angle to the coordinate directions will be less well resolved than shocks aligned with the grid. In practical calculations this is sometimes avoided by choosing grids that fit the problem well, so the shock wave is nearly aligned with the grid. This is easiest to accomplish for steady state problems, although some adaptive methods for time dependent problems also manage this.

In spite of these difficulties, many high quality computer codes have been developed that are based to a large extent on the solution of 1D Riemann problems and relatively straightforward extensions of the ideas developed in the previous chapters. It should be noted, however, that many of these methods do take into account considerably more information about the 2D structure than the above brief description indicates (see, for example, [8] or [101]). Methods are evolving in the direction of using even more information about the multidimensional behavior. One interesting possibility is to determine the direction of primary wave propagation, based on the data, rather than simply using the coordinate directions, and to employ wave propagation in other directions as well (e.g.[18], [19], [49], [69], [62]).

The development of outstanding fully multidimensional methods (and the required mathematical theory!) is one of the exciting challenges for the future in this field.

Bibliography

[1] D. A. Anderson, J. C. Tannehill, and R. H. Pletcher, *Computational Fluid Mechanics and Heat Transfer*, McGraw-Hill, 1984.

[2] J. P. Boris and D. L. Book, *Flux corrected transport I, SHASTA, a fluid transport algorithm that works*, J. Comput. Phys., 11 (1973), pp. 38–69.

[3] J. P. Boris and E. S. Oran, *Numerical Simulation of Reactive Flow*, Elsevier, 1987.

[4] L. Brillouin, *Wave Propagation and Group Velocity*, Academic Press, 1960.

[5] J. M. Burgers, *A mathematical model illustrating the theory of turbulence*, Adv. Appl. Mech., 1 (1948), pp. 171–199.

[6] A. J. Chorin, *Random choice solution of hyperbolic systems*, J. Comput. Phys., 22 (1976), pp. 517–533.

[7] J. Cole and E. M. Murman, *Calculations of plane steady transonic flows*, AIAA J., 9 (1971), pp. 114–121.

[8] P. Colella, *Multidimensional upwind methods for hyperbolic conservation laws*. Lawrence Berkeley Lab Report LBL-17023, 1984.

[9] P. Colella, *A direct Eulerian MUSCL scheme for gas dynamics*, SIAM J. Sci. Stat. Comput., 6 (1985), pp. 104–117.

[10] P. Colella and P. Woodward, *The piecewise-parabolic method (PPM) for gas-dynamical simulations*, J. Comput. Phys., 54 (1984), pp. 174–201.

[11] R. Courant and K. O. Friedrichs, *Supersonic Flow and Shock Waves*, Springer, 1948.

[12] R. Courant, K. O. Friedrichs, and H. Lewy, *Uber die partiellen Differenzengleichungen der mathematisches Physik*, Math. Ann., 100 (1928), pp. 32–74.

[13] R. Courant, K. O. Friedrichs, and H. Lewy, *On the partial difference equations of mathematical physics*, IBM Journal, 11 (1967), pp. 215–234.

[14] R. Courant, E. Isaacson, and M. Rees, *On the solution of nonlinear hyperbolic differential equations by finite differences*, Comm. Pure Appl. Math., 5 (1952), p. 243.

[15] M. G. Crandall and A. Majda, *The method of fractional steps for conservation laws*, Math. Comp., 34 (1980), pp. 285–314.

[16] M. G. Crandall and A. Majda, *Monotone difference approximations for scalar conservation laws*, Math. Comp., 34 (1980), pp. 1–21.

[17] C. M. Dafermos, *Polygonal approximations of solutions of the initial-value problem for a conservation law*, J. Math. Anal. Appl., 38 (1972), pp. 33–41.

[18] S. F. Davis, *A rotationally biased upwind difference scheme for the Euler equations*, J. Comput. Phys., 56 (1984), pp. 65–92.

[19] H. Deconinck, C. Hirsch, and J. Peuteman, *Characteristic decomposition methods for the multidimensional Euler equations*, in 10th Int. Conf. on Num. Meth. in Fluid Dyn., Springer Lecture Notes in Physics 264, 1986, pp. 216–221.

[20] R. J. DiPerna, *Finite difference schemes for conservation laws*, Comm. Pure Appl. Math., 25 (1982), pp. 379–450.

[21] B. Einfeldt, *On Godunov-type methods for gas dynamics*, SIAM J. Num. Anal., 25 (1988), pp. 294–318.

[22] B. Engquist and S. Osher, *Stable and entropy satisfying approximations for transonic flow calculations*, Math. Comp., 34 (1980), pp. 45–75.

[23] C. A. J. Fletcher, *Computational Techniques for Fluid Dynamics*, Springer-Verlag, 1988.

[24] S. K. Godunov, Mat. Sb., 47 (1959), p. 271.

[25] J. B. Goodman and R. J. LeVeque, *On the accuracy of stable schemes for 2D scalar conservation laws*, Math. Comp., 45 (1985), pp. 15–21.

[26] J. B. Goodman and R. J. LeVeque, *A geometric approach to high resolution TVD schemes*, SIAM J. Num. Anal., 25 (1988), pp. 268–284.

[27] A. Harten, *High resolution schemes for hyperbolic conservation laws*, J. Comput. Phys., 49 (1983), pp. 357–393.

[28] A. Harten, *ENO schemes with subcell resolution*, J. Comput. Phys., 83 (1987), pp. 148–184.

[29] A. Harten, B. Engquist, S. Osher, and S. Chakravarthy, *Uniformly high order accurate essentially nonoscillatory schemes, III*, J. Comput. Phys., 71 (1987), p. 231.

[30] A. Harten and J. M. Hyman, *Self-adjusting grid methods for one-dimensional hyperbolic conservation laws*, J. Comput. Phys., 50 (1983), pp. 235–269.

[31] A. Harten, J. M. Hyman, and P. D. Lax, *On finite-difference approximations and entropy conditions for shocks*, Comm. Pure Appl. Math., 29 (1976), pp. 297–322 (with appendix by Barbara Keyfitz).

[32] A. Harten and P. D. Lax, *A random choice finite difference scheme for hyperbolic conservation laws*, SIAM J. Num. Anal., 18 (1981), pp. 289–315.

[33] A. Harten, P. D. Lax, and B. van Leer, *On upstream differencing and Godunov-type schemes for hyperbolic conservation laws*, SIAM Review, 25 (1983), pp. 35–61.

[34] A. Harten and S. Osher, *Uniformly high-order accurate nonoscillatory schemes. I*, SIAM J. Num. Anal., 24 (1987), pp. 279–309.

[35] A. Harten, S. Osher, B. Engquist, and S. Chakravarthy, *Some results on uniformly high-order accurate essentially nonoscillatory schemes*, Appl. Numer. Math., 2 (1986), pp. 347–377.

[36] A. Harten and G. Zwas, *Self-adjusting hybrid schemes for shock computatons*, J. Comput. Phys., 9 (1972), p. 568.

[37] G. Hedstrom, *Models of difference schemes for $u_t + u_x = 0$ by partial differential equations*, Math. Comp., 29 (1975), pp. 969–977.

[38] C. Hirsch, *Numerical Computation of Internal and External Flows*, Wiley, 1988.

[39] A. J. Jameson, W. Schmidt, and E. Turkel, *Numerical solutions of the Euler equations by a finite-volume method using Runge-Kutta time-stepping schemes*. AIAA Paper 81-1259, 1981.

[40] G. Jennings, *Discrete shocks*, Comm. Pure Appl. Math., 27 (1974), pp. 25–37.

[41] F. John, *Partial Differential Equations*, Springer, 1971.

[42] H. O. Kreiss and J. Lorenz, *Initial-Boundary Value Problems and the Navier-Stokes Equations*, Academic Press, 1989.

[43] N. N. Kuznetsov, *Accuracy of some approximate methods for computing the weak solutions of a first-order quasi-linear equation*, USSR Comp. Math. and Math. Phys., 16 (1976), pp. 105–119.

[44] P. D. Lax, *Hyperbolic systems of conservation laws, II.*, Comm. Pure Appl. Math., 10 (1957), pp. 537–566.

[45] P. D. Lax, *Hyperbolic Systems of Conservation Laws and the Mathematical Theory of Shock Waves*, SIAM Regional Conference Series in Applied Mathematics, #11, 1972.

[46] P. D. Lax and B. Wendroff, *Systems of conservation laws*, Comm. Pure Appl. Math., 13 (1960), pp. 217–237.

[47] B. P. Leonard, *Universal limiter for transient interpolation modeling of the advective transport equations: the ULTIMATE conservative difference scheme*. NASA Technical Memorandum 100916, NASA Lewis, 1988.

[48] R. J. LeVeque, *Time-split methods for partial differential equations*, PhD thesis, Stanford, 1982.

[49] R. J. LeVeque, *High resolution finite volume methods on arbitrary grids via wave propagation*, J. Comput. Phys., 78 (1988), pp. 36–63.

[50] R. J. LeVeque and B. Temple, *Stability of Godunov's method for a class of 2×2 systems of conservation laws*, Trans. Amer. Math. Soc., 288 (1985), pp. 115–123.

[51] H. W. Liepmann and A. Roshko, *Elements of Gas Dynamics*, Wiley, 1957.

[52] J. Lighthill, *Waves in Fluids*, Cambridge University Press, 1978.

[53] T. P. Liu, *The deterministic version of the Glimm scheme*, Comm. Math. Phys., 57 (1977), pp. 135–148.

[54] B. J. Lucier, *Error bounds for the methods of Glimm, Godunov and LeVeque*, SIAM J. Num. Anal., 22 (1985), pp. 1074–1081.

[55] R. W. MacCormack, *The effects of viscosity in hypervelocity impact cratering*. AIAA Paper 69-354, 1969.

[56] A. Majda, *Compressible fluid flow and systems of conservation laws in several space variables*, Springer-Verlag, Appl. Math. Sci. Vol. 53, 1984.

[57] O. Oleinik, *Discontinuous solutions of nonlinear differential equations*, Amer. Math. Soc. Transl. Ser. 2, 26 (1957), pp. 95–172.

[58] S. Osher, *Riemann solvers, the entropy condition, and difference approximations*, SIAM J. Num. Anal., 21 (1984), pp. 217–235.

[59] S. Osher, *Convergence of generalized MUSCL schemes*, SIAM J. Num. Anal., 22 (1985), pp. 947–961.

[60] S. Osher and S. Chakravarthy, *High resolution schemes and the entropy condition*, SIAM J. Num. Anal., 21 (1984), pp. 995–984.

[61] R. Peyret and T. D. Taylor, *Computational Methods for Fluid Flow*, Springer, 1983.

[62] K. G. Powell and B. van Leer, *A genuinely multi-dimensional upwind cell-vertex scheme for the Euler equations*. AIAA Paper 89-0095, Reno, 1989.

[63] R. D. Richtmyer and K. W. Morton, *Difference Methods for Initial-value Problems*, Wiley-Interscience, 1967.

[64] P. L. Roe, *Approximate Riemann solvers, parameter vectors, and difference schemes*, J. Comput. Phys., 43 (1981), pp. 357–372.

[65] P. L. Roe, *Numerical algorithms for the linear wave equation*. Royal Aircraft Establishment Technical Report 81047, 1981.

[66] P. L. Roe, *The use of the Riemann problem in finite-difference schemes*. Lecture Notes in Physics 141, Springer-Verlag, 1981.

[67] P. L. Roe, *Some contributions to the modeling of discontinuous flows*, Lect. Notes Appl. Math., 22 (1985), pp. 163–193.

[68] P. L. Roe, *Characteristic-based schemes for the Euler equations*, Ann. Rev. Fluid Mech., 18 (1986), pp. 337–365.

[69] P. L. Roe, *Discrete models for the numerical analysis of time-dependent multidimentional gas dynamics*, J. Comput. Phys., 63 (1986), pp. 458–476.

[70] R. Sanders, *On convergence of monotone finite difference schemes with variable spatial differencing*, Math. Comp., 40 (1983), pp. 91–106.

[71] A. H. Shapiro, *The Dynamics and Thermodynamics of Compressible Fluid Flow*, Ronald Press, Cambridge, MA, 1954.

[72] C. Shu, *Total-variation-diminishing time discretizations*, SIAM J. Sci. Stat. Comput., 9 (1988), pp. 1073–1084.

[73] C. Shu and S. Osher, *Efficient implementation of essentially non-oscillatory shock capturing schemes, II*, J. Comput. Phys., to appear.

[74] C. Shu and S. Osher, *Efficient implementation of essentially non-oscillatory shock capturing schemes*, J. Comput. Phys., 77 (1988), pp. 439–471.

[75] C.-W. Shu, *TVB uniformly high-order schemes for conservation laws*, Math. Comp., 49 (1987), pp. 105–121.

[76] J. Smoller, *On the solution of the Riemann problem with general step data for an extended class of hyperbolic systems*, Mich. Math. J., 16 (1969), pp. 201–210.

[77] J. Smoller, *Shock Waves and Reaction-Diffusion Equations*, Springer, 1983.

[78] G. Sod, *A survey of several finite diffeence emthods for systems of nonlinear hyperbolic conservation laws*, J. Comput. Phys., 27 (1978), pp. 1–31.

[79] G. Sod, *Numerical Methods for Fluid Dynamics*, Cambridge University Press, 1985.

[80] G. Strang, *Accurate partial difference methods II: nonlinear problems*, Numer. Math., 6 (1964), p. 37.

[81] G. Strang, *On the construction and comparison of difference schemes*, SIAM J. Num. Anal., 5 (1968), pp. 506–517.

[82] J. C. Strikwerda, *Finite Difference Schemes and Partial Differential Equations*, Wadsworth & Brooks/Cole, 1989.

[83] P. K. Sweby, *High resolution schemes using flux limiters for hyperbolic conservation laws*, SIAM J. Num. Anal., 21 (1984), pp. 995–1011.

[84] E. Tadmor, *The large-time behavior of the scalar, genuinely nonlinear Lax-Friedrichs scheme*, Math. Comp., 43 (1984), pp. 353–368.

[85] E. Tadmor, *Numerical viscosity and the entropy condition for conservative difference schemes*, Math. Comp., 43 (1984), pp. 369–381.

[86] L. N. Trefethen, *Group velocity in finite difference schemes*, SIAM Review, 24 (1982), pp. 113–136.

[87] M. Van Dyke, *An Album of Fluid Motion*, Parabolic Press, Stanford, CA, 1982.

[88] B. van Leer, *Towards the ultimate conservative difference scheme I. The quest of monotonicity*, Springer Lecture Notes in Physics, 18 (1973), pp. 163–168.

[89] B. van Leer, *Towards the ultimate conservative difference scheme II. Monotonicity and conservation combined in a second order scheme*, J. Comput. Phys., 14 (1974), pp. 361–370.

[90] B. van Leer, *Towards the ultimate conservative difference scheme III. upstream-centered finite-difference schemes for ideal compressible flow*, J. Comput. Phys., 23 (1977), pp. 263–275.

[91] B. van Leer, *Towards the ultimate conservative difference scheme IV. A new approach to numerical convection*, J. Comput. Phys., 23 (1977), pp. 276–299.

[92] B. van Leer, *Towards the ultimate conservative difference scheme V. A second order sequel to Godunov's method*, J. Comput. Phys., 32 (1979), pp. 101–136.

[93] B. van Leer, *On the relation between the upwind-differencing schemes of Godunov, Engquist-Osher, and Roe*, SIAM J. Sci. Stat. Comput., 5 (1984), pp. 1–20.

[94] W. G. Vincenti and C. H. Kruger Jr., *Introduction to Physical Gas Dynamics*, Wiley, 1967.

[95] J. Von Neumann and R. D. Richtmyer, *A method for the numerical calculation of hydrodynamic shocks*, J. Appl. Phys., 21 (1950), pp. 232–237.

[96] R. Warming and Hyett, *The modified equation approach to the stability and accuracy analysis of finite-difference methods*, J. Comput. Phys., 14 (1974), pp. 159–179.

[97] G. Whitham, *Linear and Nonlinear Waves*, Wiley-Interscience, 1974.

[98] P. Woodward and P. Colella, *The numerical simulation of two-dimensional fluid flow with strong shocks*, J. Comput. Phys., 54 (1984), pp. 115–173.

[99] H. Yee, *A class of high-resolution explicit and implicit shock-capturing methods.* Von Karman Institute for Fluid Dynamics, Lecture Series 1989-04, 1989.

[100] H. C. Yee and A. Harten, *Implicit TVD schemes for hyperbolic conservation laws in curvilinear coordinates.* AIAA Paper 85-1513, Cincinnati, 1985.

[101] S. T. Zalesak, *Fully multidimensional flux corrected transport algorithms for fluids*, J. Comput. Phys., 31 (1979), pp. 335–362.

[102] S. T. Zalesak, *A preliminary comparision of modern shock-capturing schemes: linear advection*, in Advances in Computer Methods for Partial Differential Equations, VI, R. Vichnevetsky and R. S. Stepleman, eds., IMACS, 1987, pp. 15–22.

LeVeque, R.J.
Numerical Methods for Conservation Laws
Second edition
1994 (3rd printing 1994). 232 pages. Softcover. ISBN 3-7643-2723-5

Tromba, A.J.
Teichmüller Theory in Riemann Geometry
Based on lecture notes by J. Denzler
1992. 220 Pages. Softcover. ISBN 3-7643-2735-9

Bättig, D. / Knörrer, H.
Singularitäten
1991. 140 pages. Softcover. ISBN 3-7643-2616-6

Narasimhan, R.
Compact Riemann Surfaces
1992. (2. printing 1992). 120 pages. Softcover. ISBN 3-7643-2742-1

Nevanlinna, O.
Convergence of Iterations for Linear Equations
1993. 188 pages. Softcover. ISBN 3-7643-2865-7

Baumslag, G.
Topics in Combinatorial Group Theory
1993. 172 pages. Softcover. ISBN 3-7643-2921-1

Birkhäuser

DMV Seminar
Workshops, edited by the German Mathematics Society

Mathematics with Birkhäuser

The workshops organized by the Gesellschaft für mathematische Forschung in cooperation with the Deutsche Mathematiker–Vereinigung (German Mathematics Society) are primarily intended to introduce students and young mathematicians to current fields of research. By means of these well-organized seminars, scientists from other fields will also be introduced to new mathematical ideas. The publication of these workshops proceedings in the DMV-Seminar series will make the material available to an ever larger audience.

Birkhäuser

Made in the USA
San Bernardino, CA
25 August 2017